Preparation of Metal Nanoparticles by Plant Materials-Applications

Dr.T. Sivakumar, M.Sc., M.Phil., Ph.D.,

Assistant Professor,

Department of Botany,

Annamalai University, Annamalai Nagar,

Tamilnadu, India.

Published by

Preparation of Metal Nanoparticles by Plant Materials-Applications

ISBN 978-93-87862-10-4

Author

Dr.T. Sivakumar

Bonfring

309, 2nd Floor,

5th Street Extension, Gandhipuram,

Coimbatore-641 012.

Tamilnadu, India.

E-mail: info@bonfring.org

Website: www.bonfring.org

Phone: 0422 4213231

Dedicated to Beloved

Mother and Father

Teacher and God

Author Profile

Dr.T. Sivakumar received his Ph.D in 2006 from the Department of Botany at Annamalai University under the direction of Professor R. Panneerselvam. Then he Joined 2008 Lecturer in Botany, Faculty of Science, Annamalai University, Annamalai Nagar, Tamil Nadu India. Dr.T. Sivakumar focuses his research interest on the Plant Physiology and specific area of "Triazole as a plant growth regulators and also act as fungicides in edible tuber crops, vegetable and medicinal plants". Currently, he has worked as the different fields of traditional medicine, including natural product biosynthesis. Green synthesis of metal nanoparticles using medicinal plants like Cassia, green tea, synthesis, characterization, thermodynamic properties and applications of antimicrobial, antibacterial, antifungal, antioxidant and anticancer activity analyzed by cell line culture, anti diabetic, and antidandruff activities.

<table>
<tr><td>Chapter</td><td align="center">Contents</td><td>Page No</td></tr>
</table>

Chapter I

I	1. Introduction	1
	1.1. Nano-The Beginning	1
	1.2. Metal Nanoparticles	2
	1.3. Nanoparticle	3
	1.4. Plant Derived Synthesis of Metal Nanoparticles	3

Chapter II

II	2.1. Flowchart for Green Synthesis	5
	2.2. Green Synthesis of Nanoparticles	6
	2.3. Synthesis of Metallic Nanoparticles	6
	2.4. Chemical Reduction Method	7
	2.5. Reduction by Citrate Anion	9
	2.6. Physical Method	9
	2.7. Biological Method	10

Chapter III

III	3. Characterization of Metallic Nanoparticles	12
	3.1. UV-Vis Spectroscopy	12
	3.2. X-ray Diffraction Study	12
	3.3. FT-IR Spectroscopy Analysis	13
	3.4. FE-SEM Analysis	13
	3.5. TEM Analysis	13
	3.6. SEM Analysis	15
	3.7. Absorbance Spectroscopy	15
	3.8. AFM	15
	3.9. EXAFS	16
	3.10. XPS	16

Chapter IV

IV	4.1.	General Application of Metallic Nanoparticles	17
	4.2.	Application of Metallic Nanoparticles and Optical Function	17
	4.3.	Therapeutic Applications of Metallic Nanoparticles	19
	4.4.	Application of Agriculture	23
	4.5.	Nanoparticles Based Smart Delivery Systems	31
	4.6.	Nanostructured Formulation Reduce Nutrients Loss into Soil by Leaching and/or Leaking	36
	4.7.	Protein Polymer-Based Nanoparticles	41
	4.8.	Other Proteins: Casein, Fibrinogen, Hemoglobin, Bovine Serum Albumin, Gluten	48
	4.9.	Factors to Control Particle Formation	52

Chapter V

V	5. Novel Applications of Protein-Based Nanoparticles	55
	5.1. Bioimaging	55
	5.2. Drug Delivery Vehicle	56
	References	59

CHAPTER I

1. Introduction

Nanoscience and nanotechnology refer to the control and manipulation of matter at nanometer dimensions. This control has made it possible to have life, which is a collection of most efficient nanoscale processes. The best cost effective, eco-friendly and efficient processes must learn from nature. When we explore life around us, it is found that organization of nanomaterials is central to biology. Architectures made by organisms are all based on nanoassemblies. Today I know that it is possible to use biological processes to make artificial nanostructures. Chemically synthesized nanostructures have been used at various stages of civilization.

1.1. Nano-The Beginning

One of the earlist Nano-sized objects known to us was made of gold. Faraday prepared colloidal gold in 1856 and called it "Divided metals". In his diary dated 2April 1856, Faraday called the particles he made the divided state of gold (http://personal.bgsu.edu/~nberg/faraday/diary2.htm). Metallic gold, when divided into fine particles ranging from size of 10-500nm particles, can be suspented in water. In 1890, the German bacteriologist Robert Koch found that compounds made with gold inhibited the growth of bacteria. He won the Nobel Prize for medicine in 1905. The use of gold in medicinal preparations is not new. In the Indian medical system called Ayurveda, gold is used in several preparations. One popular preparation is called '*Saraswatharishtam*', prescribed for memory enhancement. Gold is also added in certain medicinal preparations for babies, in order to enhance their mental capability. All these preparations use finely ground gold. The metal was also used for medical purposes in ancient Egypt. Over 5,000 years ago, the Egyptians used gold in dentistry. In Alexandria, alchemists developed a powerful colloidal elixir known as 'liquid gold', a preparation that was meant to restore youth. The great alchemist and founder of modern medicine, Paracelsus, developed many highly successful treatments form metallic minerals including gold. In Chine, people cook their rice with a gold coin in order to help replenish gold in their bodies. Colloidal gold has been incorporated in glassea and vases to give them colour. The oldest of these is the fourth Century AD Lycurgus cup made by the Romans. The cup appears red in transmitted light (if a light source is kept within the cup) and appears green in reflected light (if the light source is outside). Modern chemical analysis shows that the glass is not much different from that used today [1].

However, the science of nanometer scale objects was not discussed until much later. On December 29, 1959, the Nobel prize winning physicist, Rechard Feynman gave a talk at the annual meeting of the American Physical Society entitled "There's plenty of room at the bottom'. In this talk he stated, "The principles of physics, as far as I can see, do not speak against the possibility of maneuvering things atom by atom". He in a way, suggested the bottom up approach, "it is interesting that it would be, in principle, possible (I think) for a physicst to synthesize any chemical substance that the chemist writes down [2].

1.2. Metal Nanoparticles

In medieval era, metallic nanoparticles were actually used to decorate cathedral windows. Due to unique properties of noble metal nanoparticles, it has made a special place in the field of nanotechnology. The most important feature of nanoparticles is their surface area to volume ratio, where it easily allows them to interact with other particles. In nanoparticles, high surface area to volume ratio makes diffusion faster and is feasible at lower temperatures. And this field has found more interesting, without disturbing and poisoning of healthy cells, we can directly treat affected cells and tissues. In fluorescence enhancement and surface enhanced Raman spectroscopy and in environment refractive index sensing nanoparticles have found additional application in the enhancement of field sensitive optical process. The optical properties of metal nanoparticles play a key role due to the localized surface Plasmon with resonance wavelength in the visible region. Silver and gold nanoparticles are effective in inhibiting growth of gram-positive and gram negative bacteria. For the production of nano devices, living organism has huge potential. However, it requires much more experimentation. There is a drawback such as involvement of toxic chemicals makes it difficult for synthesis of metallic nanoparticles.

So, there is an alternate way of synthesising metallic nanoparticles by using living organisms such as fungi, bacteria, plants. Several studies have shown that metallic nanoparticles characteristics like (size, stability, physical, chemical properties, morphology) are strongly influenced by the experimental conditions, adsorption process of stabilizing agent, the kinetics of interaction of metal ions with reducing agents. In various industrial applications, metallic nanoparticles have attracted, because of their different physical and chemical properties from bulk metals. Various properties like mechanical strengths, high surface area, low melting point, optical properties and magnetic properties. Catalysts which are used in metallic nanoparticles are selective and highly active, has long lifetime for many chemical reactions. It has experimented that a DVD disk with storage capacity of 10 tetra bytes, which are approximately 2000 movies of convectional size. This is possible only due to the optical properties of gold nanorods embedded in the disk which are oriented randomly. To store data, Zijlstra and team used optical spectrum and different polarization directions.

1.3.   Nanoparticle

Nanotechnology is an interesting branch of science which deals with the synthesis, characterization and applications of nanoparticles. Nanotechnology has emerged as an interdisciplinary field combining biology, physics, chemistry and material science. The production of nanoparticles can be achieved through two approaches known as 'top-down process' and 'bottom-up approach'. In the top-down approach, the bulk material is broken down into particles at nano-scale through different processes like grinding, milling, etc. In the bottom-up approach, the atoms/molecules self assemble to form new nuclei which grow into a particle of nano-scale.

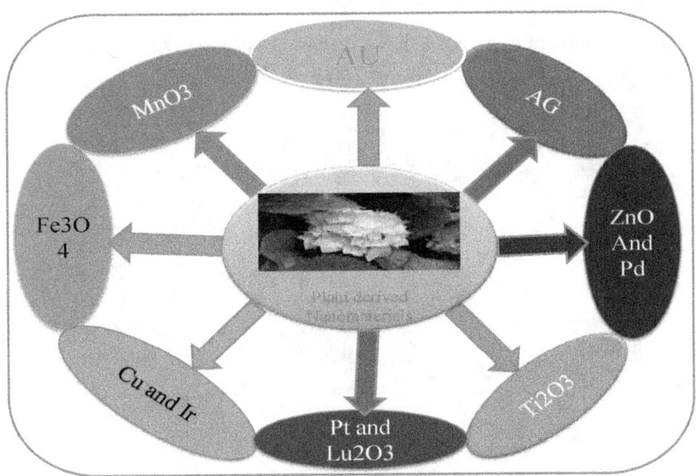

Figure 1.1: In Generally Used Methods (chemical, physical, and biological) for Synthesis of metal Nanoparticles and the Techniques Used for the Characterization of Synthesized Nanoparticles

1.4.   Plant Derived Synthesis of Metal Nanoparticles

Metal nanoparticles have been produced by physical and chemical methods for a long time, but recent developments explored the critical role of biological systems for this purpose (Figure. 1.1: 1.2). Physical and chemical methods are energy intensive processes which mean high expenditure. Production of silver nanoparticles by chemical reduction (e.g., hydrazine hydrate, sodium borohydride, DMF, and ethylene glycol) may lead to absorption of harsh chemicals on the surfaces of nanoparticles raising the toxicity issue.

Moreover, nanocrystalline silver colloids produced by such aqua-chemical routes exhibit aggregation with time, thereby compromising with the size factor upon storage. Among the bio-inspired synthesis of metals, plant extracts are found to be more suitable candidates over other biological entities (microorganisms and fungi) because they do not require toxic reducing and capping agents, radiation, high temperature, microbial/fungal strains and costly media for microbial/fungal growth as well as for nanoparticles production.

They also avoid the chances of infection/contamination during synthesis and application section. These demerits recommend the plant-mediated synthesis of metal nanoparticles which involves synthesis at biological pH. Furthermore, due to slower kinetics, it offers better manipulation and control over crystal growth and their stabilization. An economical point of view also prioritizes to plant extracts as they are ubiquitous and easily available. Besides this, the process of extract preparation is cheap and simple.

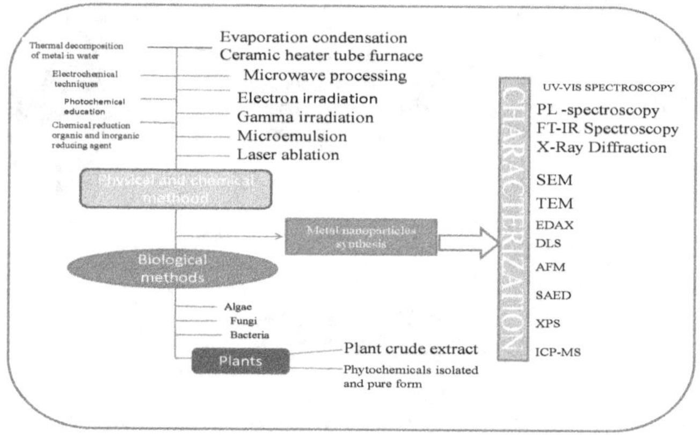

Figure 1.2: Commonly Used Methods (chemical, physical, and biological) for Synthesis of Metal Nanoparticles and the Techniques Used for the Characterization of Synthesized Nanoparticles

References

[1]   M-C. Daniel and D. Astruc, Chem. Rev., 104 (2004), pp, 293-346.

[2]   Htt: / /www.zyvex.com/nanotech/Feynman.html.

2.1. Flowchart for Green Synthesis

The general flow chart representing the green synthesis of nanoparticles and their characterization techniques is illustrated in Fig. 2.1.

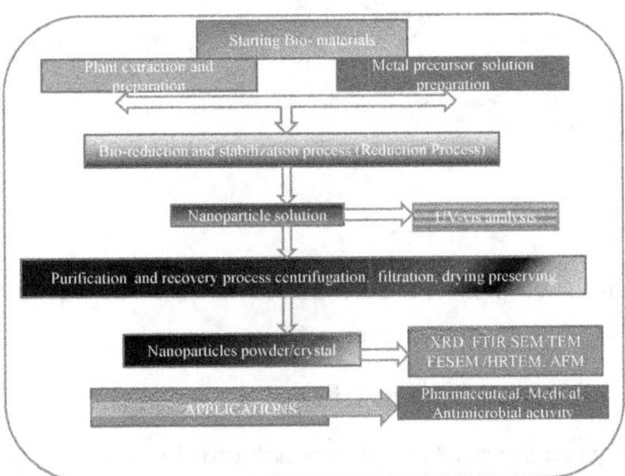

Figure 2.1: Flowchart for Green Synthesis, Characterization and Application of Nanoparticles

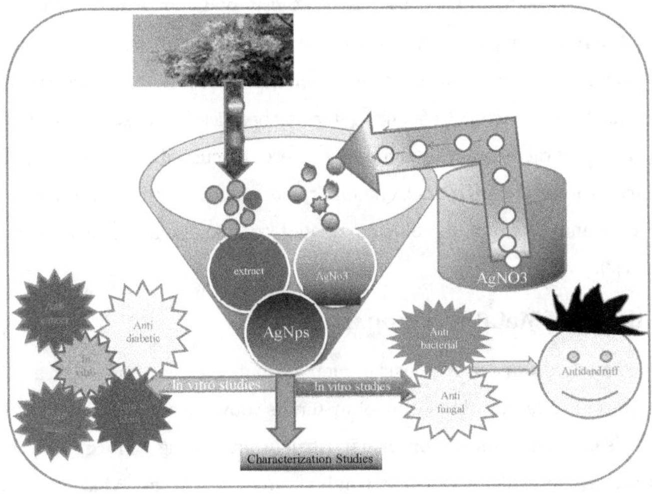

Figure 2.2: Schematic Diagram for Green Synthesis, Characterization and Application of Nanoparticles

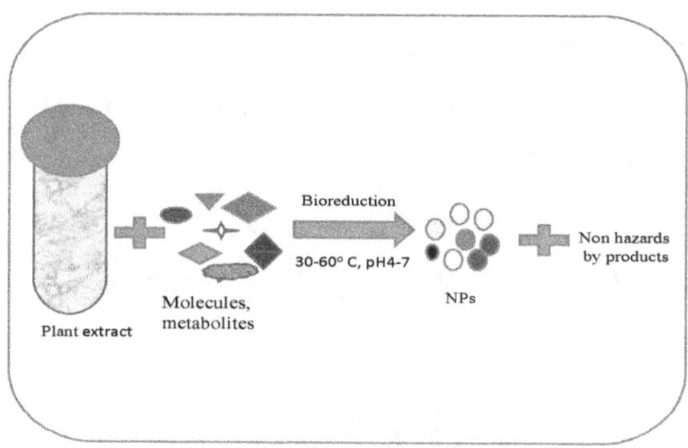

Figure 2.3: Schematic Diagram for Green Synthesis, Characterization and Applications of Nanoparticles

2.2.   Green Synthesis of Nanoparticles

The potentials of 'green chemistry' to synthesize nanoparticles not only in the laboratory scale but also in their natural environment. Factors affecting biosynthesis along with current and future applications were also presented. Researcher had been presented a review article on the role of plants as green sources towards synthesis of nanoparticles. Details regarding fabrication of different nanoparticles, probable mechanisms, applications of nanoparticles and the reviewed references were furnished. A summarizing review of literature on plants as emerging nanofactories. Factors influencing the nanoparticle synthesis, role of biomolecules, applications, and characterization techniques were discussed. Properties and potential applications are various types of metallic nanoparticles such as copper, aluminium, gold, silver, magnesium, zinc and titanium. Problems related to toxicity of nanoparticles were also addressed in brief.

2.3.   Synthesis of Metallic Nanoparticles

There is two methods for the synthesis of metallic nanoparticles- top-down and bottom-up. These approaches include the attenuation of materials components with further self-assembly process which leads to the formation of nanostructures. During self-assembly the physical forces operating at nanoscale are used to combine units into large stable structures. Typical examples include Quantum dot and formation of nanoparticles from colloidal dispersion. Top down approach- These approaches include macroscopic structures which can be externally controlled in the processing of nanostructures. Typical examples are ball milling, application of severe plastic deformation. Top down method v/s bottom up methods- Top down method starts with a pattern generated on a large scale, then reduced to nanoscale, quick to manufacture, slow and not suitable for large scale production. Bottom-up approach- begins with atoms or molecules and build up to nanostructures, fabrication is much less expensive. Attrition/milling is top-down type of method and bottom-up method is production of colloidal dispersion.

2.4. Chemical Reduction Method

2.4.1. *Gold Nanoparticles*

For centuries, gold nanoparticles have been intensively studied. When chitosan capped gold nanoparticles coupled with ampicillin, two-fold enhancement of antibacterial activity was observed [1] Chemical reduction is the most common and widely used method for the preparation of gold nanoparticles. This method includes, in the presence of reducing agent, gold salt is reduced [2]. In 1857, Michael Faraday first studied gold colloids synthesis in solution phase, where in an aqueous medium gold chloride reduced with phosphorus [3]. One article has been reported in 1951 which discovered citrate reduction method [4]. Synthesis of AuNPs was based on a single phase reduction of gold tetrachloroauric acid by sodium citrate in an aqueous medium and produced particles about 20nm. The major contribution for AuNPs synthesis was published in 1994 and now it popularly known as Burst- Schifrin method. This process use two phases that exploits thiol ligands that strongly bind to gold due to the soft character of S and Au. Firstly, with the help of a phase transfer agent such as tetraoctylammonium bromide gold salt is transferred into an organic solvent, then organic thiol is added. At last, strong reducing agent such as sodium borohydride (in excess) is added which gives thiolate protected AuNPs.

The major benefit of synthesizing this process are ease of preparation, size controlled, thermally stable nanoparticles, reduced dispersity. Natan investigated gold nanoparticles seeded growth by using modification on the Frens synthesis. Bastus have synthesized monodispersed citrate stabilized particles through kinetically controlled seed growth. Narrow size distribution was prepared with a uniform quasi spherical shape (up to 200nm) which is kinetically controlled seeded growth strategy via HAuCl4 reduction by sodium citrate. During homogeneous growth, the inhibition of any secondary nucleation was controlled by adjusting temperature, pH, and seed particle concentration. Frens method improved results in different aspects-

- It allows better control of size and size distribution of gold nanoparticles
- It produces particles of higher monodispersity
- It leads to higher concentration.

With a wide variety of molecules, this method can be further functionalized.

2.4.2. Platinum Nanoparticles

In the synthesis of platinum nanoparticles, the platinum metal precursor either in an ionic or a molecular state is taken. By the reducing agents, chemical changes are made to convert the precursor to platinum metal atoms. These metal atoms then combine into stabilizer or supported materials to form nanoparticles. For example, in chemical reduction H_2PtC_{l6} is reduced by $NaBH_4$ or Zn to give rise to platinum nanoparticles. $H_2PtC_{l6} + NaBH_4 = Pt +$ other reaction product H_2PtC_{l6} is the common precursor used in synthesis of platinum nanoparticles. H_2PtC_{l6} is usually dissolved in organic liquid phase or aqueous phase. By introducing decomposition, displacement, reducing agent, electrochemical reactions, the dissolved metal precursor can be converted into the solid metal. Radiolytic, sonochemical or electrochemical method by this three method, chemical step can be activated by physical mixing. Two different reactivities are generally used in case of mixed metal nanoparticles e.g., $RuCl_3$ and H_2PtCl_6 $\{Na_6Pt (So_3)_4, Na_6Ru (So_3)_4\}$ $\{PtCl_2$ and $RuCl_3)\}$ various complex mixed precursors have been also used.

2.4.3. Silver Nanoparticles (Ag NPs)

Five gram of plant samples are boiled in 100 mL of deionized water for 30 min, cooled to room temperature and filtered using Whatman No. 1 paper. The extract was then used for the synthesis of silver nanoparticles. 30 mg of silver nitrate ($AgNO_3$) obtained from sd Fine chemicals was dissolved in 180 mL of deionized water so as to obtain 0.1 mM of silver nitrate precursor solution. 20 mL of sprout extract was added to this solution and the mixture was magnetically stirred for 30 min. The mixture solution which initially appeared in pale green colour, slowly changed into brownish red colour within 2 h. The colour change indicated the formation of silver nanoparticles and the completion of the synthesis process. The reacted solution was centrifuged at a speed 1200 rpm for 15 min and the nanoparticles settled at the bottom were carefully removed and spread on a petridish for drying at 35°C. AgNps are one of the most attractive inorganic materials because of its environment free nature.

2.4.4. *Zinc Oxide Nanoparticles (ZnO NPs)*

Zinc acetate dihydrate with 90% purity was obtained from Himedia and distilled water was used throughout the experiments. 0.2 M of zinc acetate dihydrate was dissolved in 70 mL of distilled water and stirred for few minutes. 5 g of plant leaf powder, in dried form, was added to 100 mL of distilled water and magnetically stirred for 2 h at 80°C. After cooling to room temperature and filtering through Whatman No. 1 paper. 30 mL of this green tea extract was mixed homogenously with the already prepared zinc acetate solution. The reacted solution was dried at 60°C overnight to yield pale-white ZnO nanoparticles, which were finally calcined at 100°C for 1 h and preserved in air-tight vials for further studies.

2.5.   Reduction by Citrate Anion

From the pioneering studies; it is now well known that citrate acted in both ways. First is to stabilize the nanoparticles and to reduce the metal cation. To determine the particles growth this reactant played a major role. Citrate controls the size and shape of AgNps. This function was investigated by Pillai and kamat. At different citrate concentration, by using the boiling method, AgNps with Plasmon maximum absorbance at 420nm were produced. By increasing the concentration 1 to 5 times of sodium citrate to silver cation. i.e., [citrate] /[Ag+], the elapsed time for AgNps formation was 40 to 20min reduced respectively, which indicates that a fraction of the Ag+ was not reduced under equimolar conditions. b. Reduction by Gallic acid

At room temperature, reduction of Ag+ in water can be achieved by using Gallic acid (GA) whose oxidation potential is 0.5V. In benzoic acid structure the hydroxyl group at determined positions plays an important role in the synthesis of metal nanoparticles. When hydroxyl groups are located at Meta position, nanoparticles synthesis was not successful but it was achieved when hydroxyl groups are present at ortho and para positions. Here carboxylic group act as stabilizer and hydroxyl as the reactive part. To obtain silver colloids, NaOH addition is important. Then, the silver species reacting could be Ag2O that has been reported as a good AgNps precursor by thermal decomposition.

2.6.   Physical Method

To form AuNPs, photochemical reduction of gold salts has been used [5]. This formulation uses continuous wave UV irradiation (250-400nm), PVP as the capping agent, ethylene glycol as the reducing agent. Glycol concentration and viscosity of the solvent mixture are the two factors where AuNPs formation is dependent upon. Process was further improved by the addition of Ag+ to the solution, leading to an increase in the production of Au nanoparticles[6].

To synthesize platinum nanoparticles, irradiation and laser ablation techniques have also been used. In one method irradiation was combined with ultrasonication. So, in this process, H_2PtCl_{16} $6H_2O$ was added to a solution of 10 mm polypyrole and SDS. Particle size is controlled by varying the length and time of ultra sonication and irradiation [7].

2.7. Biological Method

Plant mediated synthesis has gained more popularity due to eco-friendly. Zingiber Officinale extract acts as a reducing agents, as well as stabilizer with particles ranging from 5-15nm in diameter. To synthesize metallic nanoparticles, several bacteria and fungus like prokaryotic bacteria and eukaryotic fungus. Plant extract may have been employed for the reduction of aqueous metal ions. Biological methods may have wide distribution in particle size but have a slow reaction rate. At room temperature, the extract is mixed with a metal salt solution, within minutes the reaction is complete. By this method, gold, silver nanoparticles have been synthesised. The rate of nanoparticles production, their quantity can be affected by the plant extract concentration, its nature metal salt concentration, temperature, the pH. By using a leaf extract of Polyalthia longifolia, silver nanoparticles were synthesized.

For preparing metallic nanoparticle by the use of plant extract is environmental friendly, economical. It brings controlled size and morphology of nanoparticles (Figure 2.4).

Figure 2.4: Various Types of Plants Used for the Synthesis of Metal Nanoparticles

References

[1] Sinha A, Manjhi J. Inter. J. Nanopart, 2015.

[2] M.A. Hayat, Colloidal gold: principles, methods, and applications, Elsevier, 2012.

[3] J.A. Khan, R.A. Kudgus, A. Szabolcs, S. Dutta, E. Wang, S. Cao, G.L. Curran, V. Shah, S. Curley, D. Mukhopadhyay and J.D. Robertson, Designing nanoconjugates to effectively target pancreatic cancer cells in vitro and in vivo, PLoS One, 6(6), 2011.

[4] J.M. Simard, Synthesis of gold nanoparticles for biomacromolecular recognition, University of Massachusetts Amherst, 2017.

[5] S. Eustis, H.Y. Hsu and M.A. El-Sayed, Gold nanoparticle formation from photochemical reduction of Au3+ by continuous excitation in colloidal solutions. A proposed molecular mechanism, The Journal of Physical Chemistry B, 109(11), Pp.4811-4815, 2005.

[6] S. Iravani, Green synthesis of metal nanoparticles using plants, Green Chemistry, Pp.2638-2650, 2011.

[7] B. De Gusseme, L. Sintubin, L. Baert, E. Thibo, T. Hennebel, G. Vermeulen, M. Uyttendaele, W. Verstraete and N. Boon, Biogenic silver for disinfection of water contaminated with viruses, Applied and Environmental Microbiology, 76(4), Pp.1082-1087, 2010.

CHAPTER III

3. Characterization of Metallic Nanoparticles

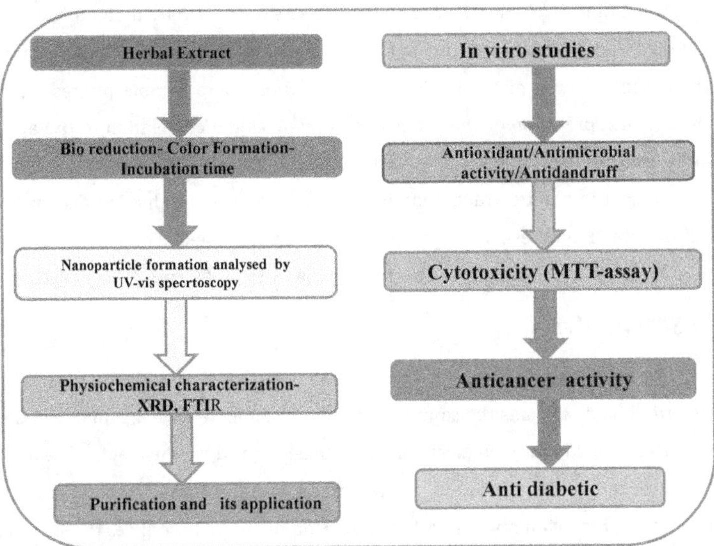

Figure 3.1: Schematic Diagram of Characterization and Applications of Metal Nanopartilces

3.1. UV-Vis Spectroscopy

UV-Vis spectrum usually depends on the size of the synthesized particle and the position of the absorption maximum shifts with the particle size. Compared to bulk materials the nanoparticles show enhanced optical property which may be due to high aspect ratio and the absorption maxima shifted towards lower wavelength region.

3.2. X-ray Diffraction Study

XRD analysis are used to find the phase purity and size of the synthesized crystallites The average crystallite size (D) of samples were calculated using the Debye-Scherrer equation.

$$D = 0.9\lambda/\beta\cos\theta$$

where λ is the wavelength of X-ray source (CuK_α line = 0.1541 nm) used in XRD, β is the full width at half maximum (FWHM) in radians and θ is the Bragg's diffraction angle X'PERT PRO X-ray diffractometer was used to record the XRD pattern of the NPs.

3.3.   FT-IR Spectroscopy Analysis

Fourier-Transform Infrared spectroscopy is a physico-chemical analytical technique that provides a snap shot of the composition of a sample. It is one of the most widely used methods for rapid identification of chemical constituents in a sample. FT-IR method measures the vibrations of the bonds within the chemical functional groups and generates a spectrum. FT-IR spectrum is a measurement of absorption of IR radiations by a sample plotted against the wavelength. The interpretation of the IR spectrum involves the correlation of the absorption bands (vibrational bands) with the chemical compounds in the sample. In this way, the biomolecules present in plant extracts that are responsible for the reduction and stabilization processes of the green synthesis of nanoparticles can be identified. The FT-IR spectra of the samples were recorded in Perkin Elmer Spectrum Version 10.03.09 using KBr pellete method.

3.4.   FE-SEM Analysis

FE-SEM provides images of a sample at magnifications 10 X to 300,000 X, with virtually unlimited depth of field. It scans the sample surface with a high-energy beam of electrons in a raster scan pattern. Field emission electron gun is used to produce the electron beam which is focused into a thin monochromatic beam. Compared to with the conventional SEM, the FE-SEM produces clearer, less distorted images with a resolution down to 1.5 nm *i.e.,* three to six times better. FE-SEM micrographic images of the NPs were recorded. Namely TEM, scanning TEM (STEM), electron energy loss spectroscopy (EELS) and x-ray photoelectron spectroscopy (XPS).

3.5.   TEM Analysis

TEM analysis (Transmission electron microscope) is also widely used to characterize nanomaterials to gain information about particle size, shape, crystallinity and interparticle interaction. TEM is a high spatial resolution structural and chemical characterization tool. It has the capability to directly image atoms in crystalline specimens at resolutions close to 0.1nm, smaller than interatomic distance. An electron beam can also be focused to a diameter smaller than ~0.3nm, allowing quantitative chemical analysis from a single nanocrystal.

TEM is a very powerful technique for the characterization of NP size, composition and crystalline structure. When an electron beam interacts with a sample, the electrons can be either transmitted, scattered, backscattered or diffracted. TEM uses the transmitted electron signal to form an image of the sample. The transmitted electron beam is dependent on the sample thickness; for thin samples (a few nanometres), the transmitted electrons pass through without significant energy loss.

STEM differs from TEM by focusing the electron beam into a narrow spot that is scanned over the sample in a raster. Because the attenuation of the electrons depends significantly on the density and thickness of the sample, the transmitted electron beam forms a 2D image of the sample. In hybrid samples, STEM imaging allows the identification of different components based on intensity variation. This intensity variation is related to the difference in the atomic numbers of each component (Z-contrast). In addition, the rastering of the beam across the sample makes it possible to couple STEM with other characterization methods such as EELS, allowing direct correlation of image and quantitative data thus obtain details regarding the chemical composition of NPs.

Figure 3.2: The Transmission Electron Microscope Used for the Characterization of Metallic NPs

General TEM analysis does not have sufficient resolution to determine the crystallinity of a nanomaterial. However, high-resolution TEM (HRTEM) can be successfully employed for the characterization of the crystallinity of a sample with atomic resolution, as well as for providing information regarding electron diffraction analysis. This approach helps in gaining insight into the ordering of atoms in a NP.

XPS is a surface-sensitive quantitative spectroscopic technique that measures the elemental composition, chemical states and electronic states of the elements within the material. XPS spectra are obtained by irradiating a material with a beam of x-rays while simultaneously measuring the kinetic energy of the electrons that escape from the top 0 to 10 nm of the material being analysed. XPS requires high vacuum ($P \sim 10^{-8}$ mbar) or ultra-high vacuum ($P < 10^{-9}$ mbar) conditions. However, when used to analyse NPs, the importance of the coverage of the NPs must be kept in mind: high coverage leads to high-quality spectra. On the other hand, to quantify the composition of the NPs, an inert transfer of the sample to the analysis chamber is necessary to avoid contamination and oxidation of the NP surface.

3.6. SEM Analysis

(Scanning Electron Microscopy) It is a powerful technique for imaging any material surface with a resolution down to about 1nm. The interaction of an incident electron beam with the specimen produces secondary electrons, with energies smaller than 50ev. SEM can give information about the purity of nanoparticle sample.

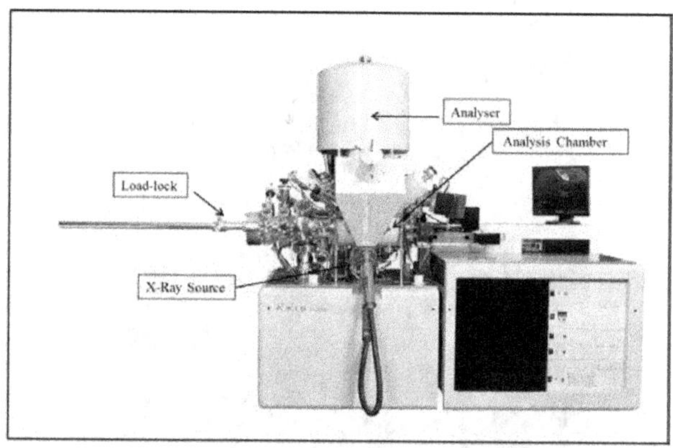

Figure 3.3: The Scanning Electron Microscope Used for the Characterization, Particle Size of Metallic Nanoparticles

3.7. Absorbance Spectroscopy

Spectroscopy is useful to characterize metal nanoparticles, because they possess bright colour which is visible by naked eye. By this technique, qualitative information about the nanoparticle can be obtained. By applying Beer's law absorbance is measured. Depending on path length (l), nanoparticle conc. (c), extinction coefficient (A) can be obtained.

3.8. AFM

It is a better choice for nonconductive nanomaterials. Typically, it has vertical resolution of less than 0.1nm and lateral resolution of around 1nm.It gives detailed information on the atomic scale, which is important for understanding electronic structure and chemical bonding of atoms and molecules.

3.9.   EXAFS

(Extended X-ray Absorption Fine Structure) this is one of the most reliable and powerful characterization technique to evaluate the structure of metallic nanoparticles; especially it is useful to determine bimetallic nanoparticles. To gain appropriate information about the structure, the sample of metallic nanoparticles should be homogeneous. This method provides the no. Of atoms surrounded the x-ray absorbing atom and their interatomic distances involved in the shells.

3.10. XPS

(X-ray Photoelectron Spectroscopy) it is used to provide information on metal state. Suppose the oxidation state of metal on the surface. It is often oxidized by air. So, by using this method, 0-valency of surface metal must be confirmed.

4.1. General Application of Metallic Nanoparticles

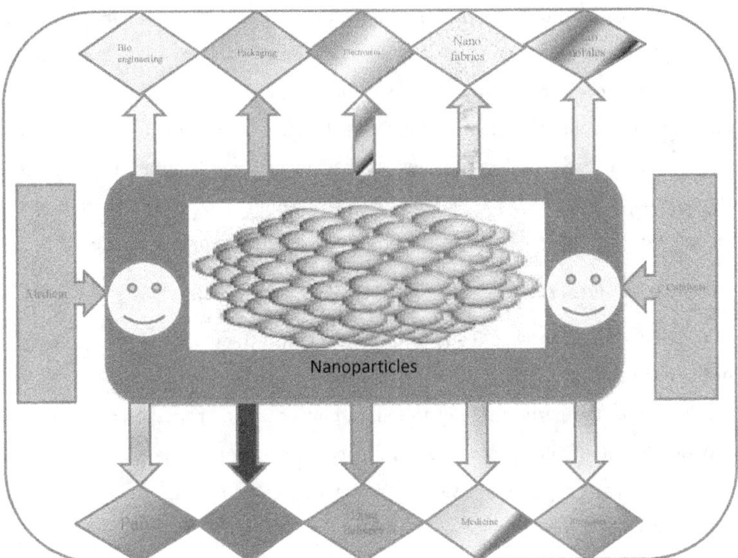

Figure 4.1: Schematic Diagram and Application of Metal Nanoparticles

4.2. Application of Metallic Nanoparticles and Optical Function

Imaging sensor, display, solar cell, Photo catalysis, biomedicine, optical detector, laser-this are the applications based on the optical properties of metal nanoparticles. It is mainly dependent on some factors such as shape, size, surface area, doping, and interaction with the surrounding. The optical properties of CdSe semiconductor nanoparticles can change with size. For different samples of gold nanospheres, the optical properties changes with enlargement of metallic nanoparticles. Surface absorption Plasmon Au & Ag can change into various colors by changing the particle size, form and shape of the particle and condensation rate.

4.2.1. Thermal Function

When nanoparticle diameter is less than 10nm, the melting point is also lower than a bulk metal. With low boiling point, electronic wiring can be made with nanoparticles.

4.2.2. Electrical Function

Can be used to make high temperature superconductivity material. In conductance one step can be shown, at constant applied voltage, the mechanical thinning of a nanowire and electric current measurement. So the main point here to be noted is that, the number of electron wave modes supporting to the electrical conductivity is becoming smaller with decreasing diameter of the wire. Only one electron wave mode is observed in electrically conducting carbon nanotubes which transport the electrical current. Electrically conducting carbon nanotubes touch the mercury surface at different times, as their length and orientation are different and this leads to transport of electrical current. This gives two type of information i) different nanotubes resistance ii) the effect of carbon nanotubes length on the resistance.

4.2.3. Mechanical Function

Polymers filled with nanotubes leads to improvement in their mechanical properties. And this progress is purely dependent on the filler type and the way with which the filing is conducted. The larger the particle size of the filler, poorer is the properties obtained. Polymer matrix and defoliated phyllosilicates consisting components provide excellent mechanical properties. Mechanical property of metallic nanoparticles can be improved by mixing the nanoparticles with metals or ceramics.

4.2.4. Magnetic Function

At the nanosized level, pt and gold nanoparticles exhibit magnetic property but as bulk they are non-magnetic. By capping, the nanoparticle surface and bulk atoms can be improvised by interaction with other chemical species. So, by capping with appropriate molecules; this gives the chances to modify the physical property of nanoparticles.

4.2.5. Catalysis

Catalysts based on metallic nanoparticles are-selective, highly active, exhibit long lifetime for several kinds of reactions. So, there are two types of catalyst- Heterogeneous catalysts- which are immobilized on inorganic support. Applications- Oxidation reductions, synthesis of H_2O_2, water gas shift, hydrogenation. Homogenous catalysts- metallic nanoparticles surrounded with stabilizers. Applications- Nitrile hydrogenation, olefin hydrogenation.

4.2.6. Used as Fuel Cell Catalysts

Fuel cell is a device that directly converts chemical potential energy into electric energy. A PEM (Proton Exchange Membrane) cell uses hydrogen gas (H_2) & oxygen gas (O_2) as fuel. The products of fuel cell are water, electricity, heat.

4.2.7.   Used in Materials Science

Nickel nanoparticles are used as electrical conductive pastes, battery materials etc.

4.2.8.   Used in Medical Treatment

Healthy cell can be distinguished from cancer cell by the presence of Antibodies joined to the Au nanoparticle.

4.2.9.   Used in Paints

Nano titanium dioxide is used in paint to exploit two outstanding properties- photo catalytic activity, UV protection. Addition of nanosilicon dioxide to paints can improve the macro and micro hardness, abrasion, scratch resistant.

4.2.10. Elimination of Pollutants

As metallic nanoparticles is highly active in terms of physical, chemical and mechanical properties. They can be used as catalysis to prevent environment pollution arising from coal and burning gasoline. As they react with toxic gases such as carbon monoxide and nitrogen oxide.

4.2.11. Used as Sun Screen Lotion

Nanomaterials are very useful as sunscreen lotions by blocking UV radiation effectively for a prolonged period of time. As prolonged UV exposure causes skin burns. By applying sun-screen lotions containing nano-TiO_2 it gives sun protection factor (SPF).

4.3.   Therapeutic Applications of Metallic Nanoparticles

4.3.1.   As Anti-Infective Agents

The anti-viral properties of AgNps are more effective then chemically synthesized silver nanoparticles [1]. In one study, metallic nanoparticles have been described as a HIV preventative therapeutic [2].

In a couple of studies, it has been shown that as virucidal agent silver acts directly on the virus by binding to the glycoprotein gp120 [3]. This binding in turn prevents the CD4 dependent virion binding which effectively decreases HIV- 1's infectivity [4] and it has also been reported that metallic nanoparticles has been effective antiviral agents against herpes simplex virus [5], influenza [6], respiratory syncytial viruses [7].

4.3.2. As anti-Angiogenic

It is well known that angiogenesis is the development of new blood vessels and occurs during normal development and in some disease states. It plays a main role in number of diseases such as cancer, rheumatoid arthritis. In normal conditions, angiogenesis is tightly regulated between various pro-angiogenic growth factors (VEGF, PDGF, and TGF-B) and anti-angiogenic factors. Under diseased conditions, angiogenic is turned on. Some reviews have reported that these agents have serious toxicities such as fatal haemorrhage, thrombosis, and hypertension. It may be overcome if these nanoparticles alone can be efficacious as an anti-angiogenic agent.

4.3.3. In Tumour Therapy

It has been studied that naked gold nanoparticles inhibited the activity of heparin-binding proteins such as VEGF165 and bFGF in vitro and VEGF induced angiogenesis in vivo [8]. Further work in this area has been reported that onto the surface of AuNPs heparin binding proteins are absorbed [9] and were subsequently denatured[10].The researchers also showed that surface size plays a main role in the therapeutic effect of AuNPs. Mukherjee and colleagues also experimented the effect of gold nanoparticles on VEGF mediated angiogenesis using a mouse ear model injected with an adrenoviral vector of VEGF (Ad-VEGF- mimics the resulting angiogenic response found in tumours)[11]. A week later, the Ad- VEGF administration, mice treated with AuNPs developed lesser edema than the same treated mice. Eom and Colleagues revealed the anti-tumour effects of 50nm AgNps In vitro and In vivo.

4.3.4. In Multiple Myeloma

Researchers (Washington university school of Medicine in St.Louis, journal Molecular cancer Therapeutics) have designed a nanoparticle based therapy that is effective in treating mice with multiple myeloma. Multiple myeloma is a cancer that affects plasma cells. Mukherjee and group demonstrated that a gold nanoparticle inhibits the VEGF and bFGF dependent proliferation of multiple myeloma cells.

4.3.5. In Leukaemia

B-chronic Lymphocytic Leukaemia (CLL) is an incurable disease predominantly characterized by apoptosis resistance, by co-culture with an anti-VEGF antibody, found induction of more apoptosis in CCL B cells. In CLL therapy, gold nanoparticles were used to increase the efficacy of these agents. Gold nanoparticles were chosen based on their biocompatibility, very high surface area, surface functionalization and ease of characterization.

To the gold nanoparticles, VEGF antibodies were attached and determined their ability to kill CLL B cells.

4.3.6. In Rheumatoid Arthritis

Scientists from the University of Wollongong (Australia) have built a new class of anti-arthritic drug which could be used by gold nanoparticles and it has fewer side effects. Rheumatoid arthritis is an autoimmune disease that occurs when the immune system not function properly and attacks a patient's joints. New research has shown that gold particles can invade macrophages, and stop them from producing inflammation without killing them. Journal of inorganic biochemistry it has been published that by reducing the size of gold into smaller nanoparticles (50nm) was able to cause more gold to immune cells with lesser toxicity.

4.3.7. In Photo Thermal Therapy

Gold nanoparticles absorb light strongly as they convert photon energy into heat quickly and efficiently. Photo-thermal therapy (PTT) is an invasive therapy in which photon energy is converted into heat to kill cancer.

4.3.8. In Radiotherapy

Tumours loaded with gold, this absorbs more X-rays as gold is an excellent absorber of X-rays. Thus deposition of more beam energy and results in a local dose which increase specifically to tumour cells. Gold nanoparticles have been more useful to treat cancer.

4.3.9. Potential Side Effects of Metal Nanoparticles

Argyria is one of the reported side effects in patients which results from prolonged contact with or ingestion of silver salts. Argyria is characterized by gray, black staining of the skin and mucous membrane produced by silver deposition. Silver may be deposited in the skin either from silver salts containing medications or from industrial exposure. One report was suggested where a patient ingested colloidal silver 3 times a year over a 2year period resulted in diabetics, hypertension, hyperlipidaemia [12]. Mouse brains exposed to nanosilver reported apoptosis and gene modulation [13]. Due to inhalation, workers also had tarnished corneas and conjunctiva [14]. Without any reported side effects, gold colloid has been used for centuries in therapeutic field. It has also been reported that gold nanoparticle causes thrombosis, immunogenic reactions, and haemolysis [15]. Enzyme present in saliva can transform gold (0) to gold (I), which is subsequently engulfed by immune cells.

Some side effects related to this are erythema nodosum, allergic reactions, macular and papular rash. While injected with gold complexes leads to very low incidence of nephrotoxicity with minor proteinuria. Due to teratogenicity of gold complexes it is not recommended for pregnant women because it can leads to haematological disorder. In healthy human body, gold is present in the range of 0-0.001ppm [16]. It is found in small amounts in skin (0.03ug/g), hair (0.3ug/g) and nails (0.17ug/g) [17-19]. It has been reported that Au, Ag and Pt nanoparticles toxicity over 72 hour period using Zebrafish model. It causes delayed hatching, crippled backbone, cardiac disorders, and platinum accumulation in brain when they found polyvinyl alcohol capped at particles (3- 10nm).It has been reported that the neurotoxicity of Cu and Mn nanoparticles in PC-12 cell lines.

4.3.10. *Metallic Nanoparticle as Drug Delivery*

Most of the chemotherapeutics agents distribute to the whole body results in toxicity and it gives poor compliance by patients, so as targeted delivery of therapeutic agents to tumour cells is a challenge. By active and passive targeting, imaging of tumour cells is done by metallic nanoparticles. Both at surface and inside cells, metallic nanoparticles can interact with bio molecules because of their small size which gives better targeting for therapeutics. Between 10-100nm [20] of different shapes, sizes of gold, nickel, silver, iron [21] metallic nanoparticles have been checked out as diagnostics and drug delivery systems. Gold nanoparticles utility in cancer cells and in xenograft tumour mouse models was experimented and reported the use of non-toxic PEG gold nanoparticles for tumour targeting (in vivo) which were biocompatible and were characterized by SERS (surface enhanced Raman scattering) [22]. But the use of metallic nanoparticles for drug delivery is a concern because after drug administration, some fraction of metallic particles can be retained in the body even though it is inert and biocompatible.

These metallic nanoparticles can be easily conjugate with various agents such as peptides, antibodies and DNA/RNA to specifically target different cells [23], with polymers (polyethylene glycol) which are biocompatible to prolong their circulation in vivo for drug and gene delivery applications [24-25]. They can also transform light into heat, thus enabling thermal ablation of targeted cancer cells [26-28]. For the delivery of anticancer drugs such as Paclitaxel [29] or cisplatin, oxiplatin (platinum based drugs); Au nanoparticles have been used as vehicles. This has investigated 2nm Au nanoparticles covalently bind with the chemotherapeutic drug paclitaxel. Gold-gold sulphide nanoshells have been produced as a photo thermal modulated drug delivery system. These nanoshells covered by a hydro gel matrix which are thermo sensitive. These nanoshells were basically designed to strongly absorb NIR light and to release multiple bursts of any soluble material held within the hydro gel matrix in response to repeated NIR irradiation. By using, 50nm hollow Au-nanocubes with eight lopped off porous corners which are covered by a thermo sensitive polymer containing preloaded effectors that can be released in a controllable manner using an NIR laser. This work was studied by Yavuz and team. Protein modified 10, 20, and 40nm Au nanoparticles and 20, 50, and 100nm Ag nanoparticles which are obtained from fetal bovine serum have potential effects on radiation induced killing of glioma cells.

4.4.   Application of Agriculture

Agriculture has always been the backbone of most of the developing countries. It does not only fill the people abdomen but also fuel the economy. According to 2014-2015 censuses, India's population is 1.27. With the concern of providing food to such a big population there has to be a new technology giving more yield in short period. In that manner, nature is complex which will have imbalances which directly affects plants and crops, and indirectly animals and human. In according to this, other factors which affect agriculture are deficiencies in macro and micro nutrient content, population explosion, industrialization, depletion of water source, difference in soil condition, and erosion of top soil. In agriculture the main reason to use fertilizer is to give full-fledged macro and micro nutrients which usually soil lacks. 35-40 per cent of the crop productivity depends upon fertilizer, but some of the fertilizer affects the plant growth directly.

To overcome all these draw backs a smarter way *i.e.,* nanotechnology can be one of the source. Since fertilizers are the main concern, developing nano based fertilizer would be a new technology in this field. Fertilizers are sprayed in many ways either to soil or through leaves, even to aquatic environments; these inorganic fertilizers are supplied in order to provide three main components, nitrogen, phosphorous and potassium in equal ratios. Nano fertilizers increases the Nutrient use efficiency (NUE) by 3 times and it also provides stress tolerating ability. Irrespective of the type of crop nono technology can be used, nano technology increase bio source use and eco friendly in nature, builds carbon uptake and improves soil aggregation. Since these nano fertilizers contain nutrients, growth promoters encapsulated in nano scale polymers, they will also have a slow and a targeted efficient release. Nanotechnology is gathering information of atom in nano scale range, with consideration to physical, catalytic, magnetic and optical properties. However, the concentration of usage chronically exposes soil microbes and micro fauna, as well as the plants themselves, to level of chemical reactivity that may be toxic. When comparing to chemical fertilizers requirement and cost, nano fertilizers are economically cheaper and are required in lesser amount. For years farmers have found that nitrogen uptake is the main reason for improper yield. In the past days development of sensing devices are in boom. When it comes to test a particular analyte from the soil causing disturbance in the filed there are assays which give accurate result but it has a drawback of consumption of time and also the high cost for performing. Sensors are those give better results with the live pictures and conditions of the field. Sensors do monitor changes or the effects caused by various pesticides, fertilizers, and herbicide, also the physical conditions of soil like pH, moisture level, and growth conditions of crop, stem fruit or even root, toxicity studies, it can constantly monitor the toxicity produced in the field.

Since it is a human friendly sensor it starts detecting and alarms farmer so as to indicate any correct measures to be taken before rather than acting for a consequence later. When it comes to wireless technology certain node installation is carried out which makes the person to monitor the happenings in the field all the nodes can be controlled at the same time through cloud computing or even through air programming. Therefore various types of new sensors and types of fertilizers are reviewed at glance to place nanotechnology at the highest level.

4.4.1. Properties of the Nanoparticles

Two principal factors cause the properties of nanomaterials to differ significantly from other materials increased relative surface area and quantum effects. Morphology-aspect ratio/ size, hydrophobicity, solubility-release of toxic species, surface area/ roughness, Surface species contaminations/adsorption, during synthesis/history, Reactive oxygen species (ROS) O_2 / H_2O, capacity to produce ROS, structure/composition, competitive binding sites with receptor and dispersion/ aggregation are the important properties of nanoparticles.

4.4.2. Top ten Applications of Nanotechnologies in the Developing Countries

Salamanca-Buentello has categorically grouped the top ten applications of the anotechnologies and the detailed examples of the applications.

4.4.3. Nanotechnology Applications in Agriculture and Food Production

The prediction is that nanotechnology will transform the entire food industry changing the way food is produced, processed, packaged, transported and consumed. This report reviews the key aspects of these transformations, highlighting current research in the agriood industry and what future impacts these may have.

4.4.4. *Nano-Agriculture*

In December 2002, the United States Department of Agriculture (USDA) drafted the world's first "roadmap" for applying nanotechnology to agriculture and food. A large group of policy makers, land grant university representatives and corporate scientists met at Cornell University (New York, USA) to share their vision on how to remake agriculture using nanoscale technologies.

Agriculture, according to the new nano-vision, needs to be more uniform, further automated, industrialized and reduced to simple functions. In our molecular future, the farm will be a wide area biofactory that can be monitored and managed from a laptop and food will be crafted from designed substances delivering nutrients and calories efficiently to the body. Nano biotechnology will increase agriculture's potential to harvest feed stocks for industrial processes. Mean while tropical agricultural commodities such as rubber, cocoa, coffee and cotton–and the small-scale farmers who grow them-will find themselves quaint and irrelevant in a new nano economy of "flexible matter" in which the properties of industrial nanoparticles can be adjusted to create cheaper, "smarter" replacements.

Just as Genetically Modified (GM) agriculture led to new levels of corporate concentration all along the food chain, so proprietary nano technology, deployed from seed to stomach, genome to gullet, will strengthen the grasp of agribusiness over global food and farming at every stage – all, ostensibly, to feed the hungry, safeguard the environment and provide consumers with more choice. For two generations, scientists have manipulated food and agriculture at the molecular level. Agro-nano connects the dots in the industrial food chain and goes one step further down. With new nanoscale techniques of mixing and harnessing genes, GM plants become atomically modified plants. Pesticides can be more precisely packaged to knock-out unwanted pests and add artificial flavourings and natural nutrients engineered to please the palate. Visions of an automated, centrally-controlled industrial agriculture can now be implemented using molecular sensors, molecular delivery systems and low-cost labour. The European Union's (EU) vision is of a "knowledge-based economy" and as part of this; it plans to maximize the potential of biotechnology for the benefit of EU economy, society and the environment. There are new challenges in this sector including a growing demand for healthy, safe food; an increasing risk of disease and threats to agricultural and fishery production from changing weather patterns. However, creating a bioeconomy is a challenging and complex process involving the convergence of different branches of science. Nano technology has the potential to revolutionize the agricultural and food industry with new tools for the molecular treatment of diseases, rapid disease detection, enhancing the ability of plants to absorb nutrients *etc.*

4.4.5. *Controlled Environment Agriculture (CEA)*

Smart sensors and smart delivery systems will help the agricultural industry combat viruses and other crop pathogens. In the near future nanostructured catalysts will be available which will increase the efficiency of pesticides and herbicides, allowing lower doses to be used. Nanotechnology will also protect the environment indirectly through the use of alternative (renewable) energy supplies and filters or catalysts to reduce pollution and clean-up existing pollutants. An agricultural methodology widely used in the USA, Europe and Japan, which efficiently utilises modern technology for crop management, is called Controlled environment agriculture (CEA). CEA is an advanced and intensive form of hydroponically-based agriculture. Plants are grown within a controlled environment so that production practices can be optimized. The computerized system monitors and regulates localised environments for crops.

CEA technology, as it exists today, provides an excellent platform for the introduction of nanotechnology to agriculture. With many of the monitoring and control systems already in place, nanotechnological devices for CEA that provide "scouting" capabilities could tremendously improve the grower's ability to determine the best time of harvest for the crop, the vitality of the crop, and food security issues, such as microbial or chemical contamination.

4.4.6. *Precision Farming*

Precision farming has been a long-desired goal to maximize output (*i.e.* crop yields) while minimising input (*i.e.* fertilisers, pesticides, herbicides, etc) through monitoring environmental variables and applying targeted action. Precision farming makes use of computers, global satellite positioning systems, and remote sensing devices to measure highly localized environmental conditions thus determining whether crops are growing at maximum efficiency or precisely identifying the nature and location of problems. By using centralised data to determine soil conditions and plant development, seeding, fertilizer, chemical and water use can be fine-tuned to lower production costs and potentially increase production- all benefiting the farmer.

In the Erosion Technology and Concentration (ETC) group down to farm it is described that precision farming can also help to reduce agricultural waste and thus keep environmental pollution to a minimum. Although not fully implemented yet, tiny sensors and monitoring systems enabled by nanotechnology will have a large impact on future Precision farming methodologies. One of the major roles of nanotechnology-enabled devices will be increased use of autonomous sensors linked into a GPS system for real-time monitoring. These nanosensors could be distributed throughout the field where they can monitor soil conditions and crop growth. Wireless sensors are already being used in certain parts of the USA and Australia. For example, one of the Californian vineyards, Pick berry, in Sonoma county has installed wifi systems with the help of the IT company, Accenture. The initial cost of setting up such a system is justified by the fact that it enables the best grapes to be grown which in turn produce finer wines, which command a premium price. The use of such wireless networks is of course not restricted to vineyards, for example, Forbes Magazine has reported that small nanosensors are being used by Honeywell (a technology R&D company with global branches) to monitor grocery stores in Minnesota. This technology enables shop keepers to identify food items which have passed their expiry date and also reminds them to issue a new purchase order. The global market for wireless sensors is predicted to be Seven billion USD by 2010. The union of biotechnology and nanotechnology in sensors will create equipment of increased sensitivity, allowing an earlier response to environmental changes. For example:

- Nanosensors utilising carbon nanotubes12 or nanocantilevers 13 are small enough to trap and measure individual proteins or even small molecules.
- Nanoparticles or nanosurfaces can be engineered to trigger an electrical or chemical. Signal in the presence of a contaminant such as bacteria.
- Other nanosensors work by triggering an enzymatic reaction or by using nanoengineered branching molecules called dendrimers as probes to bind to target chemicals and proteins. Ultimately, precision farming, with the help of smart sensors, will allow enhanced productivity in agriculture by providing accurate information, thus helping farmers to make better decisions.

4.4.7. Encapsulating Control

Nanotechnology enables companies to manipulate the properties of the outer shell of a capsule in order to control the release of the substance to be delivered. 'Controlled release' strategies are highly prized in medicine since they can allow drugs to be absorbed more slowly, at a specific location in the body or at the say-so of an external trigger. With potential applications across the food chain (in pesticides, vaccines, veterinary medicine and nutritionally-enhanced food), these nano- and micro-formulations are being developed and patented by agribusiness and food corporations such as Monsanto, Syngenta and Kraft. Examples of nano and microcapsule designs are:

- **Slow release** – the capsule releases its payload slowly over a longer period of time (*e.g.,* for slow delivery of a substance in the body).

- **Quick-release** – the capsule shell breaks upon contact with a surface (*e.g.,* when pesticide hits a leaf)

- **Specific release** – the shell is designed to break open when a molecular receptor binds to a specific chemical (*e.g.,* upon encountering a tumour or protein in the body)

- **Moisture release** – the shell breaks down and releases contents in the presence of water *(e.g.,* in soil)

- **Heat-release** – the shell releases ingredients only when the environment warms above a certain temperature

- **pH release** – nanocapsule breaks up only in specific acid or alkaline environment (*e.g.,* in the stomach or inside a cell)

- **Ultrasound release** – the capsule is ruptured by an external ultrasound frequency

- **Magnetic release** – a magnetic particle in the capsule ruptures the shell when exposed to a magnetic field

- **DNA nanocapsule** – the capsule smuggles a short strand of foreign DNA into a living cell which, once released, hijacks cell machinery to express a specific protein (used for DNA vaccines)

4.4.8. *Nano Sensors*

Nanotechnology applications are also being developed to improve soil fertility and crop production. Nano - sensors could also monitor crop and animal health and magnetic nano-particles could remove soil contaminants. "Lab on a chip" technology also could have significant impacts on developing nations.

4.4.9. *Nanotechnology and Food Systems*

Since food systems encompass food availability, access and utilization, the scope of applications of nanotechnologyfor enhancing food security must encompass entire agricultural production - consumption systems. Further, in a rapidly globalizing economy, increasing access to food and its utilization in rural areas will be determined primarily by increase in rural incomes. The primary source of increasing rural incomes has been recognized as value addition across the different links in the agricultural production consumption chain[30-31]. These links include farm inputs, farm production systems, post *J. Farm Sci.,29(1): 2016* harvest management and processing and ûnally markets and consumers. From the food security perspective, it is therefore necessary that application of nanotechnology be not limited to the farm production level, but be extended across all the links of the agricultural value chain to increase agricultural productivities, product quality, consumer acceptance and resource use efficiencies. This will help to reduce farm costs, raise the value of production, increase rural incomes and enhance the quality of the natural resource base of agricultural production systems. In doing so, it is important to view nanotechnology as an enabling technology that can complement conventional technologies and biotechnology. Considering the concerns on biosafety and consumer acceptance emerging after agribiotechnology based products have entered the market place during last two decades, it is also essential that integrating and deploying new technologies like nanotechnology in agricultural and food systems be made after understanding the various societal and environmental implications[32,33].

4.4.10. Assessing Nanotechnology for Enhanced Food Security in India

Table 4.4.1: Application of Nanotechnology in Agriculture

Applications	nanoparticles	Reference
A)Pesticide delivery Chemical		
Avermectin	Porous hollow silica (15nm)	[34]
Ethiprole or phenylpyrazole	Poly-caprola ctone(135 nm)	[35]
Gamma cyhalothrin	Solid lipid (300 nm)	[36]
Tebucanazole/chlorothalonil	Polyvi nylpyridine andpolyvinylp yridine-co-styrene(100 nm)	[37]
Biopesticides		
Plant origin: nanosilica for insectcontrol	Nanosilic a (3-5 nm)	[38]
Essential oil encapsulation	Solid lipid (294 nm)	[39]
Microganisms: Lagenidiumgiganteum cells in emulsion	Silica (7-14 nm)	[40]
Microbial product: absorption of Myrothrecium verrucaria enzyme	Chitos an/kaolin (250-350 nm)	[41]
B). Fertilizer delivery		
NPK controlled delivery	Nano-coating of sulfur (100 nm layer) Chitos an (78 nm)	[42]
Genetic materia l deliveryDNA	Gold (10-15 nm) Gold (5-25 nm) Starch (50-100 nm)	[43], [44]
Double stranded RNA	hitosan (100-200 nm)	[45]
C). Pesticide sensor		
Carbofuran /triazophos	Gold (40 nm)	[46]
DDT	Gold (30 nm)	[47]
Dimethoate	Iron oxide (30 nm),zirconium oxide (31.5 nm)	[48]
Organophosphate	Zirconium oxide (50 nm)	[49]
Paraoxon	Paraoxon Silica (100-500 nm) Carbon	[50]
Pyrethroid	Iron oxide (22 nm)	[51]
Pesticide degradation Lindane	Iron sulûde (200 nm)	[52]
Imidacloprid	Titanium oxide (30 nm)	[53]

Assessment of emerging technologies like nanotechnology is difficult because historical data is not available for impact assessment and much of the work is at basic research stage with future promise of a range of applications. In such situations, bibliometrics and patent analysis can be used to both assess current status and trends in technology development and classify and map them to relevant application areas for strategic planning. A general premise is that basic research is found largely in journals; where as potential commercial applications are found in patents. Patent documents are also well structured to provide standardized information about citation, issue date, inventors, institutions and their locations technology ûeld classiûcation *etc.* Such structured documentation makes them suitable for assessing technology developments in various areas. Bibliometric data on the other hand is less precisely structured but amenable to formal key word searches and more intensive text mining approaches for technology assessment. A holistic systems framework was developed for patent and bibliometric analysis for assessment of the potential of nanotechnology for enhancing food systems security in India.

The frame-work was developed in two stages:

i. Mapping nanotechnology to agri-food thematic areas across all the links of the agricultural value chain (that is, farm inputs, production systems, post harvest management including storing and processing, markets and consumption).

ii. Mapping nanotechnology to the determinants of food security (productivity, soil health, water security and food quality). Vandana reviewed that potential applications of nanotechnology in agriculture are: delivery of nanocides– pesticides encapsulated in nanomaterials for controlled release;stabilization of biopesticides with nanomaterials; slow release of nanomaterial assisted fertilizers, biofertilizers and micronutrients for efficient use and ûeld applications of agrochemicals, nanomaterials assisted delivery of genetic material for crop improvement. Nanosensors for plant pathogen and pesticide detection, and NPs for soil conservation or remediation are other areas in agriculture that can beneût from nanotechnology. Enzyme immobilization for nanobiosensor using nanomaterials involves the high value low volume application of enzymes. Usually, costly, large enzyme volumes are required for biocontrol in agricultural ûelds that would be practical if spray applications combined high volume with low value. Cost-effectiveness of such biocontrol preparations can be achieved by immobilization of enzyme/ inhibitors on nanostructures, providing large surface areas, to increase the effective concentration of the preparation.

4.5. Nanoparticles Based Smart Delivery Systems

4.5.1. Applications and Advantages

Nanoscale devices have the capability to detect and treat an infection, nutrient deficiency, or other health problem, long before symptoms were evident at the macro-scale. Smart delivery systems have the capacity to monitor the effects of the delivery of pharmaceuticals, nutraceutical, nutrients, food supplements, bioactive compounds, probiotics, chemicals, insecticides, fungicides, vaccinations. Delivery systems in agriculture are important for application of pesticides and fertilizers as well as during genetic material mediated plant improvement. Application systems for pesticides need to focus on efficacy enhancement and spray drift management while fertilizers face problems of bioavailability due to soil chelation, over-application and run-offs. A viable alternative for these problems is provided by controlled delivery systems for the pesticide and fertilizer application. Controlled delivery technique aims towards measured release of necessary and sufficient amounts of agrochemicals over a period of time, to obtain the fullest biological efficacy and to minimize the harmful effects (Tsuji, 2001). For this purpose micronic and sub-micronic particles were explored as agrochemical delivery vehicles. In comparison to micronic particles (^{3}1000 nm), nano particles (b 1000 nm) offered the advantage of effective loading due to the larger surface area, easy attachment and fast mass transfer. Controlled release of the active ingredient is achieved due to the slow release characteristics of the nanomaterial, bonding of the ingredients to the material and the environmental conditions. In case of genetic material, delivery systems face challenges such as limited host range, transportation across cell membrane and trafficking to the nucleus. Nevertheless, the use of NPs assisted delivery for genetic material to develop insect resistant plant varieties is being studied. For instance, DNA-coated gold NPs are used as bullets in 'gene gun' system, for bombardment of plant cells and tissues to achieve gene transfer (Vijayakumar *et al.*, 2010)

4.5.2. Delivery of Pesticides/Biopesticides

Nanotechnology has the potential for efficient delivery of chemical and biological pesticides using nanosized preparations or nanomaterial based agrochemical formulations. The beneûts of nano material based formulations are the improvement of efficacy due to higher surface area, higher solubility, induction of systemic activity due to smaller particle size and higher mobility and lower toxicity due to elimination of organic solvents in comparison to conventionally used pesticides and their formulations. In case of biopesticides, NPs can play a major role in enhancing the efficacy and stability of whole cells, enzyme and other natural products used. However, in the ûeld, application of NPs for the delivery of pesticides and biopesticides face several challenges such as multiple environmental perturbations, large areas under spray coverage and ûnally cost effectiveness. In the usual spraying regime the whole crop is sprayed with the chemical for the ease of application involving a high volume, low value preparation. Whereas nanomaterial based preparations are expected to involve a low volume, high value applications. Such controlled nanoparticulate deli very systems will require a targeted delivery approach focused using the knowledge of the life-cycle and the behaviour of the pathogen or pest.

4.5.3. Delivery of Fertilizers

Localized application of large amounts of fertilizer, in the form of ammonium salts, urea, and nitrate or phosphate compounds are harmful. Besides much of the fertilizers are unavailable to plants as they are lost as run-off leaching causing pollution (Wilson *et al.,* 2008). Nanomaterials have potential contributions in slow release of fertilizers. Nanocoatings or surface coatings of nanomaterials, on fertilizer particles hold the material more strongly from the plant due to higher surface tension than conventional surfaces. Moreover, nanocoatings provide surface protection for larger particles. Fertilizers with sulfur nanocoating (£ 100 nm layer) are useful slow release fertilizers as the sulfur contents are beneûcial especially for sulfur deûcient soils. The stability of the coating reduced the rate of dissolution of the fertilizer and allowed slow sustained release of sulphur coated fertilizer. In addition to sulphur nanocoatings or encapsulation of urea and phosphate and their release will be beneûcial to meet the soil and crop demands. Other nanomaterials with potential application include kaolin and polymeric biocompatible NPs used biodegradable, polymeric chitosan NPs (~ 78 nm) for controlled release of the NPK fertilizer sources such as urea, calcium phosphate and potassium chloride.

4.5.4. Field Application of Nano Pesticides and Fertilizers

The mode of pesticide and fertilizer application inûuences their efficiency and environmental impact. Currently spraying of pesticides involves either knapsacks that deliver large droplets (9-66 ìm) associated with splash loss or ultra light volume sprayers for controlled droplet application with smaller droplets (3-28 ìm) causing spray drift. Constraints due to droplet size may be overcome by using NP encapsulated or nanosized pesticides that will contribute to efficient spraying and reduction of spray drift and splash losses. nother practical problem faced during pesticide application in the ûeld is settlement of formulation components in the spray tank and clogging of spray nozzles. The recent nano-sized fungicide

(~ 100 nm, BannerM AXX, Syngent) prevented spray tank ûlters from clogging, did not required mixing and did not settle down in the spray tank due to the smaller sized particles. Further, more this fungicide did not separate from water for up to one year due to nano-size, whereas fungicides that contained larger particle size ingredients typically required agitation every two hours to prevent clogging in the tank. Proper application method of optimum quantities of fertilizer maximized nutrient uptake and reduced pollution.

The choice of fertilizer application method mainly depends on: soil, crop, irrigation type and the nutrient applied. Current practices involving broadcasting, banding, side dressing and dusting face problems of run-off due to dissolution in soil moisture and leaching. Placement of large amounts of fertilizer near the seeds and reduction in soil moisture caused salt damage. The effect of different methods of nitrogen fertilizer application on the algal flora and biological nitrogen fixation in wetland rice soil was studied in pot and field experiments. In the broadcast method of urea application, nitrogen fixation was inhibited while growth of green algae was favored. In contrast, deep placement of urea granules (1- 2 g) did not suppress the growth of nitrogen fixing blue green algae (Roger *et al.*, 1980). Placement NP coated fertilizers may contribute to slow release of the fertilizer preventing the rapid dissolution and therefore harm to the environment.

4.5.5. *Conventional Fertilizers v/s Nano-fertilizers*

Conventional fertilizers are generally applied on the crops by either spraying or broadcasting. However, one of the major factors that decide the mode of application is the final concentration of the fertilizers reaching to the plant. In practical scenario, very less concentration to the targeted site due to leaching of chemicals, drift, runoff, evaporation, hydrolysis by soil moisture and photolytic and microbial degradation. It has been estimated that around 40-70% of nitrogen, 80-90% of phosphorus, and 50-90% of potassium content of applied fertilizers are lost in the environment and could not reach the plant which causes sustainable and economic losses. These problems have initiated repeated use of fertilizer and pesticide which adversely affects the inherent nutrient balance of the soil. According to an estimate by International Fertilizer Industry Association, world fertilizer consumption sharply rebounded in 2009-2010 and 2010-2011 with growth rates of 5-6% in both campaigns. World demand is projected to reach 192.8 Mt by 2016-2017[54]. But the large-scale use of chemicals as fertilizers and pesticides has resulted in environmental pollution affecting normal flora and fauna. Tilman [55] reported that excess use of fertilizers and pesticide increases pathogen and pest resistance, reduces soil microflora, diminishes nitrogen fixation, contributes to bioaccumulation of pesticides, and destroys habitat for birds. Hence, it is very important to optimize the use of chemical fertilization to fulfill the crop nutrient requirements and to minimize the risk of environmental pollution. Accordingly, it can be favorable that other methods of fertilization be also tested and used to provide necessary nutrients for plant growth and yield production, while keeping the soil structure in good shape and the environment clean. Nanotechnology has provided the feasibility of exploring nanoscale or nanostructured materials as fertilizer carrier or controlled-release vectors for building of the so-called smart fertilizers as new facilities to enhance the nutrient use efficiency and reduce the cost of environmental pollution. A nano-fertilizer refers to a product in nanometer regime that delivers nutrients to crops. For example, encapsulation inside nanomaterials coated with a thin protective polymer film or in the form of particles or emulsions of nanoscale dimensions. Surface coatings of nanomaterials on fertilizer particles hold the material more strongly due to higher surface tension than the conventional surfaces and thus help in controlled release Delivery of agrochemical substance such as fertilizer supplying macro and micronutrients to the plants is an important aspect of application of nanotechnology in agriculture.

4.5.6. *Nanotechnology in Crop Nutrition*

Fertilizers have played a pivotal role in enhancing the food grain production in India especially after the introduction of high yielding and fertilizer responsive crop varieties during the green revolution era. Despite the resounding success in grain yield, it has been observed that yields of many crops have begun to stagnate as a consequence of imbalanced fertilization and decline in organic matter content of soils. Excessive use of nitrogenous fertilizer affects the groundwater and also causes eutrophication in aquatic ecosystems. A disturbing fact is that the fertilizer use efficiency is 20-50 per cent for nitrogen and 10-25 per cent for phosphorus. With nano-fertilizers emerging as alternatives to conventional fertilizers, build-up of nutrients in soils and thereby eutrophication and drinking water contamination may be eliminated. In fact, nano-technology has opened up new opportunities to improve nutrient use efficiency and minimize costs of environmental protection. It has helped to divulge to recent findings that, plant roots and microorganisms can directly lift nutrient ions from solid phase of minerals. Slow-release of nano-fertilizers and nanocomposites are excellent alternatives to soluble fertilizers.

Nutrients are released at a slower rate throughout the crop growth; plants are able to take up most of the nutrients without any waste. Slow release of nutrients in the environments could be achieved by using zeolites that are a group of naturally occurring minerals having a honeycomb-like layered crystal structure. Its network of interconnected tunnels and cages can be loaded with nitrogen and potassium, combined with other slowly dissolving ingredients containing phosphorous, calcium and a complete suite of minor and trace nutrients. Zeolite acts as a reservoir for nutrients that are slowly released "on demand." Fertilizer particles can be coated with nanomembranes that facilitate slow and steady release of nutrients. Nano-fertilizer technology is very innovative but scantily reported in the literature.

Nano-fertilizer technology is very innovative but scantily reported in the literature.

However, some of the reports and patents strongly suggest that there is a vast scope for the formulation of nano-fertilizers. Significant increases in yields have been observed due to foliar application of nano particles as fertilizer. It was shown that 640 mg ha-1 foliar application (40 ppm concentration) of nanophosphorus gave 80 kg ha-1 P equivalent yield of clusterbean and pearl millet under arid environment. Currently, research is underway to develop nano-composites to supply all the required essential nutrients in suitable proportion through smart delivery system. Preliminary results suggest that balanced fertilization may be achieved through nanotechnology. Indeed the metabolic assimilation within the plant biomass of the metals, *e.g.*, micronutrients, applied as Nano-formulations through soil-borne and foliar application or otherwise needs to be ascertained.

Further, the Nano-composites being contemplated to supply all the nutrients in right proportions through the "Smart" delivery systems also needs to be examined closely. Currently, the nutrient use efficiency is low due to the loss of 50-70% of the nitrogen supplied in conventional fertilizers. New nutrient delivery systems that exploit the porous nanoscale parts of plants could reduce nitrogen loss by increasing plant uptake. Fertilizers encapsulated in nanoparticles will increase the uptake of nutrients. In the next generation of nanofertilizers, the release of the nutrients can be triggered by an environmental condition or simply released at desired specific time.Coating and cementing of nano and subnano-composites are capable of regulating the release of nutrients from the fertlizer capsule. A patented nano-composite consists of N, P, K, micronutrients, mannose and amino acids that increase the uptake and utilization of nutrients by grain crops has been reported. Research on the controlled release pattern of nutrients using clay nanoparticle is on the go at Tamil Nadu Agricultural University (TNAU) India. Adhikari has opined that nanofertilizers for slow release and efficient use of water and fertilizers by plants are the prime potential application of the technology. Encapsulation of fertilizers within a nanoparticle is one of these new facilities which are done in three ways a) the nutrient can be encapsulated inside nanoporous materials, b) coated with thin polymer film, or c) delivered as particle or emulsions of nanoscales dimensions. In addition, nanofertilizers will combine nanodevices in order to synchronize the release of fertilizer-N and -P with their uptake by crops, so preventing undesirable nutrient losses to soil, water and air via direct internalization by crops, and avoiding the interaction of nutrients with soil, microorganisms, water, and air.

4.6. Nanostructured Formulation Reduce Nutrients Loss into Soil by Leaching and/or Leaking

In China, the development of nanobased slow or controlledrelease fertilizers have been actively implemented since the beginning of this century and supported by the National High-Tech R&D Program. Significant progress has been made especially on film-coating urea and granular compound fertilizers. Some nanobased agrochemicals have been commercialized. The solubility and dispersion of insoluble mineral micronutrients; phosphate fertilizers have been significantly improved by nanosized or nanostructured processing. Kalpana Shastry have reviewed the patents and confirm that china is having the maximum number of the patents registered globally. The patented products include the nanoencapsulated products, nanofertilizers. Slow release of nutrients in the environments could be achieved by using zeolites that are a group of naturally occurring minerals having a honeycomb-like layered crystal structure. Its network of interconnected tunnels and cages can be loaded with nitrogen and potassium, combined with other slowly dissolving ingredients containing phosphorous, calcium and a complete suite of minor and trace nutrients. Zeolite acts as a reservoir for nutrients that are slowly released "on demand." Fertilizer particles can be coated with nanomembranes that facilitate slow and steady release of nutrients. Rahale working on "Nutrient Release Pattern of Nano- Fertilizer Formulations" for her thesis at TNAU, India has observed intake of the nutrients by the crop to an extent of 72 per cent with nanofertilizer and only 42 per cent urea and release of nano-fertilizer where zeolite was used went on for almost 50 days while that of the urea stopped after 12 days where zeolite was used as carrier.

One of the advantages of nanoscale delivery vehicles in agronomic applications is its improved stability of the payloads against degradation in the environment, thereby increasing its effectiveness while reducing the amount applied. Nano-clay particles like substances used will reduces the porosity of the polymer, or otherwise obstructs the diffusion of the active material being released, thereby increasing the length of the path of the diffusion through the host polymer. Research on the controlled release pattern of nutrients using clay nanoparticle is on the go at Tamil Nadu Agricultural University, Coimbatore.

4.6.1. *Application of Nanotechnology in Seed Science*

Seed is nature's nano-gift to man. It is self perpetuating biological entity that is able to survive in harsh environment on its own. Nanotechnology can be used to harness the full potential of seed. Seed production is a tedious process especially in wind pollinated crops. Detecting pollen load that will cause contamination is a sure method to ensure genetic purity. Pollen flight is determined by air temperature, humidity, wind velocity and pollen production of the crop. Use of bionanosensors specific to contaminating pollen can help alert the possible contamination and thus reduces contamination. The same method can also be used to prevent pollen from

Genetically, modified crop from contaminating field crops. Novel genes are being incorporated into /seeds and sold in the market. Tracking of sold seeds could be done with the help of nanobarcodes that are encodable, machine - readable, durable and sub-micron sized taggants. Disease spread through seeds and many times stored seeds are killed by pathogens. Nano-coating of seeds using elemental forms of Zn, Mn, Pa, Pt, Au, Ag will not only protect seeds but used in far less quantities than done today. Su and Li developed a technique known as quantum dots (QDs) as a flourescence marker coupled with immuno-magnetic separation for *E coli* 0157:H7, which will be useful to separate unviable and infected seeds.

Technologies such as encapsulation and controlled release methods have revolutionized the use of pesticides and herbicides. Seeds can also be imbibed with nanoencapsulations with specific bacterial strain termed as Smart seed. It will thus reduce seed rate, ensure right field stand and improved crop performance. A smart seed can be programmed to germinate when adequate moisture is available that can be dispersed over a mountain range for reforestation. Coating seeds with nano membrane, which senses the availability of water and allow seeds to imbibe only when time is right for germination, aerial broadcating of seeds embedded with magnetic particle, detecting the moisture content during storage to take appropriate measure to reduce the damage and use of bioanalytical nanosensors to determine ageing of seeds are some possible thrust areas of research.

A group of research workers is currently working on metal oxide nano-particles and carbon nanotube to improve the germination of rainfed crops. Khodakovskaya have reported the use of carbon nanotube for improving the germination of tomato seeds through better permeation of moisture. Their data show that carbon nanotubes (CNTs) serve as new pores for water permeation by penetration of seed coat and act as a passage to channelize the water from the substrate into the seeds. These processes facilitate germination which can be exploited in rainfed agricultural system.

4.6.2. Nano-herbicide for Effective Weed Control

Multi-species approach with single herbicide in the cropped environment resulted in poor control and herbicide resistance. Continuous exposure of plant community having mild susceptibility to herbicide in one season and different herbicide in other season develops resistance in due course and become uncontrollable through chemicals. Developing a target specific herbicide molecule encapsulated with nanoparticle is aimed at specific receptor in the roots of target weeds, which enter into roots system and trans located to parts that inhibit glycolysis of food reserve in the root system. This will make the specific weed plant to starve for food and gets killed.

In rainfed areas, application of herbicides with insufficient soil moisture may lead to loss as vapour. Still we are unable to predict the rainfall very preciously; herbicides cannot be applied in advance anticipating rainfall. The controlled release of encapsulated herbicides is expected to take care of the competing weeds with crops. Adjuvants for herbicide application are currently available that claim to include nanomaterials. One nanosurfactant based on soybean micelles has been reported to make glyphosate-resistant crops susceptible to glyphosate when it is applied with the 'nanotechnology-derived surfactant'. Silva in a study prepared alginate/chitosan nanoparticles as a carrier system for the herbicide paraquat. The preparation and physico-chemical characterization of the nanoparticles was followed by evaluation of zeta potential, pH, size, and polydispersion. The techniques employed included transmission electron microscopy; differential scanning calorimetry and Fourier transform infrared spectroscopy. Significant differences between the release profiles of free paraquat and the herbicide associated with the alginate/chitosan nanoparticles were observed in the study. The results showed that association of paraquat with alginate/chitosan nanoparticles alters the release profile of the herbicide, as well as its interaction with the soil, indicating that this system could be an effective means of reducing negative impacts caused by paraquat. Tests showed that soil sorption of paraquat, either free or associated with the nanoparticles, was dependent on the quantity of organic matter present.

4.6.3. Detoxification of Herbicide Residues

Excessive use of herbicides leave residue in the soil and cause damage to the succeeding crops. Continuous use of single herbicide leads to evolution of herbicide resistant weed species and shift in weed flora. Atrazine, an s-triazine-ring herbicide, is used globally for the control of pre-and postemergence broadleaf and grassy weeds, which has high persistence (half life-125 days) and mobility in some types of soils. Residual problems due to the application of atrazine herbicide pose a threat towards widespread use of herbicide and limit the choice of crops in rotation. Recent finding from TNAU, India raises hope to remediate the atrazine residue from soil within a short span of time. Application of silver modified with nanoparticles of magnetite stabilized with Carboxy Methyl Cellulose (CMC) nanoparticles recorded 88% degradation of herbicide atrazine residue under controlled environment.

4.6.4. *Nano Pesticide*

Persistence of pesticides in the initial stage of crop growth helps in bringing down the pest population below the economic threshold level and to have an effective control for a longer period. Hence, the use of active ingredients in the applied surface remains one of the most cost-effective and versatile means of controlling insect pests. In order to protect the active ingredient from the adverse environmental conditions and to promote persistence, a nanotechnology approach, namely "nano-encapsulation" can be used to improve the insecticidal value. Nanoencapsulation comprises nano-sized particles of the active ingredients being sealed by a thin-walled sac or shell (protective coating). Nanoencapsulation of insecticides, fungicides or nematicides will help in producing a formulation which offers effective control of pests while preventing accumulation of residues in soil. In order to protect the active ingredient from degradation and to increase persistence, a nanotechnology approach of "controlled release of the active ingredient" may be used to improve effectiveness of the formulation that may greatly decrease amount of pesticide input and associated environmental hazards. Nano-pesticides will reduce the rate of application because the quantity of product actually being effective is at least 10-15 times smaller than that applied with classical formulations, hence a much smaller than the normal amount could be required to have much better and prolonged management.

Several pesticide manufacturers are developing pesticides encapsulated in nanoparticles. These pesticides may be time released or released upon the occurrence of an environmental trigger (for example, temperature, humidity, light). It is unclear whether these pesticide products will be commercially available in the shortterm. Clay nanotubes (halloysite) have been developed as carriers of pesticides at low cost, for extended release and better contact with plants, and they will reduce the amount of pesticides by 70-80%, thereby reducing the cost of pesticide with minimum impact on water streams.

4.6.5. *Post Harvest Food Processing*

Food poisoning outbreaks take lives of a large number of people and also cause losses to economy in terms of lost man days and health care expenditure. Nano-materials help to keep products fresh for a longer period of time by using nanosensors placed in food production and distribution facilities, food packaging or the food itself which can detect all kinds of food pathogens like *E.coli*, Campylobactor and Salmonella by attaching themselves to the pathogens. A single nanosensor can have thousands of nano-particles that can detect the presence of any number and kind of bacteria and pathogens rapidly and accurately. Nano-sensors can work by different methods e.g. nano-sensors can be tailor-made to fluoresce into different colours or can be made out of magnetic materials.

In food and beverage industry, attempts have been made to add micronutrients and antioxidants to food substances. But these antioxidants degrade during manufacturing and food storage. Nano cocohleates delivery system protects these substances from degradation. Polyphenols and resveratrol are the substances present in most foods and wine, respectively. They get degraded and oxidized when exposed to air. Nanocochleates solve early oxidation by individually capturing and wrapping them in a phospholipids wrap and maintaining the internal nutrients secure from water and oxygen. Bio delivery Sciences International have developed nanocochleates, which are 50 nm coiled nanoparticles and can be used to deliver nutrients such as vitamins, lycopene and omega 3 fatty acids more efficiently to cells, without affecting the colour or taste of food. The delivery vehicle is made of soyphosphatidylserine which is 100% safe and provides a protective coat for range of nutrient additives.

4.6.6. *Food Packaging*

Consumers demand food to be fresh for long time, and the packaging materials should be ease for handling, safe and healthy. A major problem in food science is determining and developing an effective packaging material. Using nanoparticle technology, Bayer has developed an even more airtight plastic packaging that will keep food fresher and longer than plastics, which is "hybrid system" as it is enriched with an enormous number of silicate nanoparticles. Researchers at Leeds University have demonstrated that nanoparticles with antimicrobial properties can be employed for safer food packaging.

Nanoparticles such as titanium dioxide, zinc oxide and magnesium oxide, as well as a combination of them, once functionalized can be efficient in killing micro-organisms and are cheaper and safer instead of metal based nanoparticles. The most problematic element for food packaging engineers is oxygen because it spoils the fat in meat and cheese and turns them pale. Nanoparticles in Durethan®, Bayer's new plastic material, cannot permit air to penetrate like in other conventional plastics. Incorporation of nanoparticles of clay into an ethylene-vinyl alcohol copolymer and into a poly (lactic acid) biopolymer was found to increase barrier properties to oxygen. Polymer-silicate nanocomposites have also been reported to have improved gas barrier properties, mechanical strength, and thermal stability. Nanoclay-nylon coatings and silicon oxide barriers for glass bottles are used to impede gas diffusion.

4.6.7. *Future of Farming: Nanobio-Farming*

It is important to note that nanomaterials, owing to their increased contact surface area, might have toxic effects that are not apparent in the bulk materials especially in open agricultural ecosystems. The selection of nanomaterial for application in the field may be critical as materials which are non -toxic, biocompatible and biodegradable are desirable. Few food and nutrition products that contain nanoscale additives already in the market, such as iron in nutritional drink mixes, micelles that carry vitamins, minerals and phyto chemicals in oil and zinc oxide in breakfast cereals. The future if foreseen would allow the advancements in agricultural nanotechnology to promote 'precision farming' allowing optimum use o f the natural resources with judicious farming practices. Different sensor and controlled delivery technologies would change the face of farming. The use of a net work of sensors, global positioning system, global information system and actuators throughout an agricultural area could measure (data and statistics) and report on a number of different environmental, crop and pest variables. These reports would support the experience of the farmer and permit choices for intervention during irrigation, fertilization, pest control and even harvesting. Although costly, this would be largely offset by the rising cost of food, the need for higher quality and increasing legislation. The technology already exists to measure each of these variables. However, measurement requires technical expertise is labor-intensive and can take days, by which time the opportunity for optimal intervention could be missed. By providing robust, portable or remote in situ nanotechnology based sensing and monitoring, backed up with analytical software, farmers can begin to market her own informed choices, in real-time, and apply agrochemicals or engage expert help only when necessary.

4.7. Protein Polymer-Based Nanoparticles

4.7.1. *Fabrication and Medical Applications*

Figure 4.7.1 Nanoparticle materials can be fabricated from a variety of protein sources, including Gliadin and legumin, elastin, corn, soybean, milkprotein, albumin, gelatin etc. These proteins can then be processed into particles with unique properties for biomedical applications.

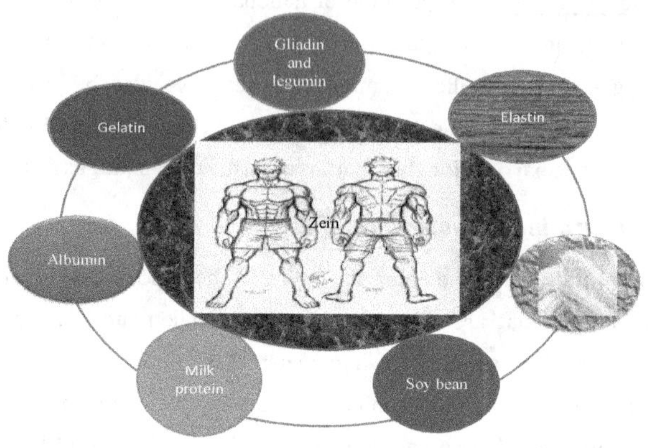

Figure 4.7.1

Drug delivery systems are a valuable means of disease treatment and prevention in today's medicine. Prior to the introduction of drug particle microencapsulation in the 1950s, drug delivery was based on rudimentary practices such as applying poultices or consuming herbal ingredients. These methods, while moderately effective at the time, are inefficient and pose unnecessary health risks. However, progress in the understanding of pharmacokinetics has led to the development of sophisticated and novel methods for administering a variety of therapeutics throughout the body. Now, drug delivery methods allow for controllable drug-release and targeting to improve the safety and efficacy of treatment. To further enhance drug delivery, nanotechnology has begun to be implemented in the field. Specifically, the use of nanoparticles as carriers is an effective strategy to deploy medications to specifically targeted parts of the body. Nanoparticles, or microspheres, are ideal drug delivery systems for both controlled and targeted drug delivery. Their sizes typically range between 1–100 nanometers in diameter and can extend to more than 1000 nanometers. However, particle sizes smaller than 200 nm are more preferable for use in nanomedicine due to their ability to traverse micro-capillaries. There are several aspects to consider, such as size and surface charge as examples, before selecting an appropriate nanoparticle material. While nanoparticles can be fabricated from synthetic materials, protein-based nanoparticles have received considerable more attention due to their biodegradability and tunable properties. Advances in medical technology have also brought about techniques to synthesize protein-based materials that offer improved efficacy and reduced costs compared to synthetic materials. Protein polymers are natural macromolecules derived from plants and animals which makes them an easily obtainable, renewable resource. In addition to their biodegradability and tunable properties, nanoparticles fabricated from protein-based materials are often biocompatible and can be easily processed [56,57]. There are a variety of different protein polymers suitable for nanoparticle-based drug delivery each with their own unique structure-function relationships. In this review, the structure and property relationships of these natural protein-based polymers will be discussed, as well as their methods of preparation. The use of these nanoparticles in medicine will then be reviewed with a focus on their application for nanoparticle-based drug delivery.

4.7.2. Categories of Protein Materials

Due to the wide range of applications for protein nanoparticles, there are many types of proteins that are used to create protein nanoparticles. The type of protein polymer required may vary depending on the application. In this book, Gliadin and legumin [58], collagen [60], gelatin [61], elastin [62], corn zein [63], and soy protein [64] will be given particular attention due to their popularity in biomaterials research (Figure 4.7.1). However, additional protein polymers such as casein (milk protein) [65], fibrinogen [66], hemoglobin [67], and have also been used to create nanoparticles.

4.7.3. Silk Fibroin

Silk fibroin protein is among the most popular natural polymers used for the creation of biomaterials due to its acceptance by the US Food and Drug Administration (FDA), low cost, and abundance. Commonly extracted from silk produced by the Bombyx mori silkworm, fibroin can be easily isolated after removal of the external sericin protein coating through treatment with sodium carbonate. The resulting fibroin protein is made of semi-crystalline structures comprised of a light and heavy chain. An isoelectric point (IEP) below pH 7 and molecular weight of 83 kDa have been reported for regenerated silk fibroin, but the latter value may vary depending on the extraction procedure and duration of treatment. The repetition of amino acids in the pattern (Gly-Ser-Gly-Ala-Gly-Ala)n leads to crystalline beta-sheets that are then stacked in an antiparallel configuration. This structure gives silk fibroin robust mechanical properties and high tensile strength. The crystallinity and conformation of silk fibroin can also be modulated to allow for high encapsulation of drugs while preserving their pharmaceutical activity.

Silk-based nanoparticles have proven effective in the delivery of both hydrophobic and hydrophilic drugs such as indomethacin and aspirin of varying molecular weights, as well as anti-cancer therapeutics such as doxorubicin, bioactive molecules including growth factors VEGF ((vascular endothelial growth factor) and BMP-2 (bone morphogenetic protein 2) and

Horseradish peroxidase and glucose oxidase enzymes, as well as plasmid DNA. Silk-composite nanoparticles have also been fabricated by combining the protein with other biopolymers such as insulin, chitosan, and albumin and synthetic polymers such as polyvinyl alcohol, polylactic acid, and polycaprolactone]. These approaches allow for a greater degree of tenability that can potentially increase the efficacy of drug delivery.

4.7.4. *Keratin*

The use of keratin as a biomaterial has been rapidly expanding over the past 40 years because of its abundance, low cost, biocompatibility, and its ability to biodegrade safely. Keratin is a fibrous structural protein with molecular weight of up to 63 kDa and IEP between pH 4.5 and 5 that is derived from the human or animal epidermis and epidermal appendages, such as hair, scales, feathers, and quills in mammals, reptiles, and birds. The keratin protein is most commonly found in epithelial cells. It is a structural protein that provides the framework for cell-cell adhesion to form a protective layer. Keratin structure is a left-handed alpha-helix which can be coiled together with other keratin proteins to form a polymerized complex. There are three different forms of keratin: α-, β-, and γ-keratins. α -keratins contain intermediate filaments, which are involved in the cytoskeleton, and are mainly found in soft tissues. β-keratins also contain intermediate filaments, but are found in hard tissues, such as scales and nails. γ -keratin is not involved in the structural elements of the cytoskeleton.

According to recent studies, keratin-based nanoparticles are effective anticancer drug carriers possessing a degree of tumor targeting ability and controlled drug release. Disulfide bonds from cysteine residues and hydrogen bonds from amine groups grant keratin nanoparticles the durability to deliver drugs with high molecular weight to their target location. In addition, keratin is negatively charged allowing positively charged molecules to better adhere to the nanoparticle for more effective transport.

The targeting ability of keratin-based nanoparticles is attributed to their pH sensitivity. Keratin-based nanoparticles can respond to changes in pH to release their drug contents accordingly in a controlled release. Due to its intrinsic water stability, keratin is also a desirable support polymer for synthetic nanoparticle composites. Silver nanoparticles coated with keratin are shown to have improved stability in aqueous environments. Keratin is also advantageous for supporting cell adhesion and promoting cellular proliferation. Gold nanoparticles coated with keratin are shown to exhibit biocompatibility with improved antibacterial activity. Keratin appears to be an ideal drug carrier which should be investigated further for drug delivery purposes.

4.7.5. *Collagen and Gelatin*

Collagen is the most abundant biopolymer in the human body. This fibrous protein is a major component of the extracellular matrix and is responsible for maintaining its structure. The majority of collagen is located in connective tissues such as the skin, tendons, and ligaments. Collagen can be divided into two different groups: non-fibrillar and fibrillar, which can be further divided depending on the structure and use. Type 1 of fibrillar collagen is the most common type found in the human body and has a molecular weight in the 100 kDa range. Long, triple helical structures are responsible for strength and flexibility in collagen. This helical structure has high mechanical strength due to a repeating amino acid sequence Gly-X-Y, where "X" and "Y" are commonly proline, hydroxyproline, leucine, or lysine. The individual helical structures, known as tropocollagens, will bind together and form a fibril structure. These fibril structures can then be cross linked together to form suitable cell scaffolds for use in tissue engineering. Due to collagen's biocompatibility and low antigenicity, collagen-based nanoparticles have been used for the delivery of pharmaceuticals such as theophylline, retinol, tretinoin, and lidocaine.

Collagen is capable of resembling themicroenvironment of some tumors allowing collagen nanoparticles to effectively infiltrate the areas and deliver anticancer therapeutics. Physical properties of collagen nanoparticles such as size, surface area, and absorption capacity, are easy to configure. This makes collagen nanoparticles a prime candidate for controlled drug release strategies. In comparison, gelatin is a biopolymer derived primarily from insoluble Type I collagen through thermal denaturation or disintegration. Like collagen, gelatin has received much attention inthe biomedical field due to its biocompatibility and high abundance. Gelatin contains a triple helical structure, similar to collagen, made of repeating amino acids: alanine, glycine, and proline. Depending on the production process, gelatin can be classified as type A or type B and consist of varying molecular weights. Type A gelatin is extracted through an acidic process, while type B is process under alkaline conditions. Type A gelatin is positively charged and has an IEP of approximately pH 9. Conversely, type B gelatin is negatively charged and has an IEP of pH 5. Tissue engineering scaffolds have thus been made from gelatin as well. Alternatively, gelatin can also be formed into a gel which can be used in the place of thermoplastic polymers. Gelatin is also a favorable nanoparticle material due to its relatively low antigenicity and non-carcinogenic nature.

Gelatin nanoparticles are extensively used as successful anticancer drug carriers and gene delivery vehicles. Gelatin nanoparticles are able to deliver drugs across the blood brain barrier, which is a semipermeable barrier that is highly studied for drug delivery systems. Gelatin nanoparticles have also safely and efficiently carried NS2, a recombinant gene from the hepatitis C virus, without negatively impacting the function of the gene. In addition, gelatin can be blended with other natural polymers to enhance their therapeutic behavior. An alginate-gelatin composite nanoparticle benefitted from an electrostatic bond formed between the two polymers and allowed for a more controlled release of the encapsulated drug, doxorubicin. Utilizing gelatin is both a promising and convenient approach for nanoparticle-based delivery of genes, vaccines, and drugs.

4.7.6. Elastin

Elastin is an important protein found in elastic fibers, specifically in the extracellular matrix. It provides support and elasticity to many structures such as the heart, lungs, skin, and blood vessels with high molecular weight species weighing 130 to 140 kDa. It is insoluble and therefore can retain its shape and insolubility after stretching. However, insoluble proteins are often not biocompatible and are difficult to alter. Through the use of recombinant proteins and peptide synthesis, soluble proteins that have elastin-like properties called elastin-like-peptides (ELP) are able to be produced with tunable molecular weights. These polypeptides are derived from tropoelastin, the building block of elastin. This precursor molecule is vital in the exploitation of ELP's. ELP's are able to react to stimuli due to their temperature sensitivity, which induces a phase transition. They can then self-assemble by the process of coacervation into a more ordered structure such as beta-spiral structure. This property, along with their biocompatibility, makes them excellent prospects for biomedical applications. Elastin-based proteins also have the ability to communicate with cells through naturally occurring cellular receptors such as elastin binding protein (EBP).

This receptor can be exploited by using tropoelastin-based polymers to induce or inhibit various cell functions. Elastin, or ELP, nanoparticles have proven effective in delivery of cytokines such as BMP-2 and -14, anticancer therapeutics such as doxorubicin, and genes. Their ability to self-assembly when exposed to certain temperatures serves as a mechanism to entrap active substances and achieve controlled drug release. The polymer functionality of ELP nanoparticles can be controlled by using a recombinant fabrication technique. This means that variables pertaining to drug release, such as composition and molecular weight, can be tailored for a variety of applications in drug delivery.

4.7.7. Corn Zein

Zein is low molecular weight protein (20 kDa), found within the cytoplasm of corn cell endosperm and is insoluble in water except in the presence of alcohol, urea, alkali, and anionic detergents. The protein has an IEP of pH 6.2 and is a mixture of two different peptides: α-zein and β- zein. α-zein is the most widely used variety due to its abundance. Zein has a helical wheel shaped structure with nine homologous units arranged in a non-parallel way with hydrogen bonds stabilizing it. This helical shape gives zein a globular structure similar to insulin and ribonuclease. Zein can be extracted using primary, secondary, and ternary solvents. Primary solvents consist of a compound that dissolves zein in a concentration greater than 10%. Secondary solvents are organic compounds. Ternary solvents are a combination of solvent, water, and alcohol. Zein is commonly used in fibers, adhesives, plastics, ink, chewing gum, and as a preservative coating for some food and pharmaceuticals. Zein nanoparticles are successful drug carriers for encapsulation and controlled release of fat soluble compounds such as α-tocopherol, other proteins, vaccines, and vitamins such as D3. Due to the protein's hydrophobicity, zein nanoparticles are promising oral drug delivery vehicles able to protect encapsulated contents from harsh acidic environments such as in stomach acid. Zein nanoparticles can also have their properties improved by combining the natural polymer with other substances. For example, sodium caseinate was incorporated with zein nanoparticles to improve particle stability in water. Zein nanoparticles are an attractive drug delivery system due to their high stability in a variety of environments and tunable properties in combination with certain molecules.

4.7.8. Soy

Soy protein is a globular protein isolated from soybeans, known as soy protein isolate, and is one of the most abundant types of plant proteins. The globular structure is comprised of two major subunits, conglycinin and glycinin, which contain all amino acids particularly glutamate, aspartate, and leucine. This structure composition gives soy protein relative stability for long storage life and biocompatibility. When the globular protein is treated with enzymes, soy protein hydrolysates below 1 kDa and between 1 and 5 kDa can be obtained and further processed. In addition, soy protein is biodegradable as it can be digested if consumed. For example, soy protein-based edible films are often used as a wax coating for fruits to preserve their quality. Soy protein films, scaffolds, and hydrogels have also been applied in tissue engineering for wound healing and transdermal drug delivery. With every amino acid available, soy protein is effective in supporting cellular communication and cell proliferation. The amino acid composition may also attribute to soy protein being used as protection against bacterial infection. Soy protein nanoparticles are becoming more popular due to the high abundance and low cost of the protein, as well as its biodegradability and low immunogenicity. The amino acid composition gives soy protein nanoparticles an advantage in encapsulation of highly hydrophobic drugs. Unlike zein, soy protein nanoparticles are soluble in aqueous environments which can be used in different oral drug delivery scenarios. Soy protein isolates are used as a coating in conjunction with other materials either for protection or for physical or chemical surface modification. For example, magnetic nanoparticles prepared with soy protein isolate benefit from enhanced functional surface area increasing the loading of enzymes. The protein coating also offers a degree of bioinert behavior to otherwise non-immunogenic nanoparticle materials.

4.8. Other Proteins: Casein, Fibrinogen, Hemoglobin, Bovine Serum Albumin, Gluten

Along with the many proteins mentioned above, there are some that will be excluded from this review but are worth mentioning. Casein, fibrinogen, hemoglobin, bovine serum albumin, and gluten are just a few of many. Similar to those previously explained, the use of these proteins depends on their properties and the application's demands. Casein is very useful in hydrophilic environments since casein is a hydrophilic protein in itself. It is useful for water-based environments since as a microsphere they disperse instead of aggregate. As a micro-/nanosphere, fibrinogen polymerizes when used in conjunction with a serine protease and forms a protein mesh that can be used to cover and treat open wounds or used in vitro for more in depth biomedical applications. Hemoglobin as a micro-/nanoparticle can be used as an oxygen deposit to make oxygen releasing biomaterials. Bovine serum albumin can be used to pack prepared protein particles to aid in protein and drug delivery. Gluten as a microsphere can be used as a drug delivery vehicle that is very effective compared to other widely used proteins. While these proteins are not described in further detail in this review, each protein possesses their own unique advantages when applied in nanoparticle-based drug delivery.

4.8.1. Fabrication Methods

Due to the necessity of obtaining particles of different sizes, shapes and weights, there are many fabrication methods that are available for the creation of nanoparticles. Fabrication methods will also vary depending on the properties of the individual polymers, such as temperature dependence. Fabrication methods that will be discussed in this review include pH variation, spray-drying, phase separation, milling, rapid laminar jet, and polymer chain collapse. The synthesis of blended protein-based nanoparticles will also be discussed. The advantages and disadvantages of these fabrication methods are summarized in Table 1.

Table 1: Advantages and Disadvantaged of the Common Protein-based Nanoparticle Fabrication Methods

Method	Advantages	Disadvantages
pH Variation	Control for particle size Control secondary structure of protein Control for zeta potential Produces chemically and physically stable particles Experimentally simple	Post-fabrication drug loading is required Limited to small scale production
Spray-drying	Cost effective Experimentally simple Easily encapsulate hydrophilic drugs Useful for heat-sensitive samples Control for particle size	Limited to small scale production Challenging to incorporate hydrophobic drugs
Rapid Laminar Jet	Control for particle size Production of uniform particles Production of strong, stable particles	Possibility of coalescence Many parameters must be controlled for
Phase Separation	Specialized equipment is not required Particle size can be controlled by adjusting protein concentration Uniform particles are produced	Particle sizes are limited to 50–500 nm in diameter Organic solvents are required Limited to small scale production
Milling	Cost effective Large scale production is possible Control of nanoparticle size Experimentally simple	Heat is released during the process requiring chamber to be cooled Little control over nanoparticle shape Nanoparticles must be coarse
Polymer Chain Collapse	Properties of the nanoparticle can be easily controlled by selection of the precursor chain Production of particles with high stability Particles with improved spherical shape are produced	Particle size is limited to 5–20 nm in diameter Side reaction may be difficult to control

4.8.2. pH Variation

The drug delivery properties of silk fibroin can be modified by changing many factors during nanoparticle synthesis. One of these factors is the pH of the silk fibroin. Particles are made by salting out a fibroin solution with potassium phosphate. The pH of the particles can be controlled depending on what type of potassium phosphate is used in the salting out. Mono potassium phosphate has a pH of 4 and dibasic potassium phosphate has a pH of 9. Silk fibroin particles with a pH of 4 develop silk II rich secondary structures while silk fibroin particles with a pH of 9 developed a silk I rich secondary structure. Particles with the silk II structure or the lower pH are less chemically stable than the particles with a higher pH and the silk I structure. When a positively charged drug is loaded into a negatively charged silk fibroin particle there is a difference in the release depending on the pH of the particle. Particle with the silk II structure and low pH have an increased initial release, whereas the high pH particles have a low release. However, particles at a neutral pH of 7 had an overall increased release over the entire time not just initially.

4.8.3. Spray-Drying

Spray-drying is a technique that is used to fabricate nanoparticles from a liquid sample. The liquid sample is sprayed out of a nozzle into a chamber where heated nitrogen and carbon dioxide gas flow in the direction of the spray. In the bottom of the chamber, there are electrodes which are used to collect the nanoparticles. As the sprayed droplets move towards the bottom of the chamber, they become electrostatically charged due to these electrodes. This is a one-step process that is a quick, cost effective method for small scale protein particle production. One application of spray-drying is for use in drug delivery systems due to the ability of hydrophilic drugs to be encapsulated in these spray-dried nanoparticles. This nanoparticle fabrication technique is useful for samples that are heat-sensitive since the solvent evaporation helps maintain the temperature of the nanoparticle droplets. This method of nanoparticle synthesis also gives the user the ability to control the size of the particle that is produced by changing parameters, such as the size of the nozzle and speed at which they are sprayed out.

4.8.4. Rapid Laminar Jet

Particles can be also made using a rapid laminar jet method. The feed liquid will contain a certain number of compounds from which the particle can be made. Spherical drops form when a liquid jet discharges from a small opening at laminar flow conditions. This formation behavior of the drops is resulted because of the surface energy and tension of the jet.

The liquid spheres will be dispersed in some type of fluid or gas/air depending on the mechanism. Drop size is-based on the length of the jet. For best results the jet length between breakpoints should be five times the diameter of the stream which gives particle sizes of about twice that of the jet. This laminar breakup of the jet is a result of small disturbances. These disturbances must be controlled to preserve uniformity in drop sizes. To control these disturbances a controlled uniform vibration is applied to the jet. Ideally the frequency of the vibration is close to the naturally occurring frequency for laminar breakup. This leads to a clean controlled breakup and uniform drop sizes.

4.8.5. Phase Separation

Out of the various methods of protein nanoparticle fabrication, emulsion-solvent evaporation is the most popular. This technique was the first to form polymer nanoparticles. Organic and aqueous phase separations are the backbone of this method. Prepared polymers are placed in an organic solvent. A surfactant is added to the aqueous phase in order to prevent the fusion of emulsion particles. The solution is then subjected to a mixing method such as ultrasonification.

Mini-emulsion droplets of polymer are formed. Finally, the solvent is separated. This is often completed by evaporation of the organic phase. The remaining solution contains polymer nanoparticles which can be collected through a centrifuge. This method produces particles in the 50–500 nm size range. Particle size could be controlled by altering the concentration of polymer solution. This technique is extremely popular due to the availability of conjugated polymers. Another method based on separations is the coacervation method. This is commonly referred to simply as phase separation but for the purposes of this paper it is included in this section. This method requires the separation of two liquid solutions. One will contain the protein polymer and the other is a solvent. Through some means of disrupting equilibrium such as the addition of a salt, coacervation is induced. The charges create electrostatic forces which induce the formation of nanoparticles.

4.8.6. Milling

Milling is a fabrication technique for nanoparticles that requires mechanical energy to break down larger particles into fine nanoparticles. This fabrication technique is commonly used for nanoparticles that are to be used in drug delivery. Milling is a cost-effective way to produce nanoparticles in a large-scale production. High energy ball milling involves the subjection of coarse nanoparticles to high energy collisions from the milling balls. Coarse nanoparticle powder is placed in a chamber that contains milling balls and mechanical movement is applied to the cylindrical chamber to accelerate the milling balls which can roll over and collide with the powder. These collisions and other mechanical force from the milling balls causes the coarse nanoparticles to break down into fine nanoparticles. The chamber must be cooled due to the heat energy released from the mechanical energy exerted on the nanoparticles. This fabrication technique allows the user of the system to control the size of the nanoparticles by altering the speed of the rotation of the cylindrical chamber.

4.8.7. Polymer Chain Collapse

Single-chain collapse of polymers is a method to produce individual single-chain polymer nanoparticles (SCNP). This method can produce particles in the range of 5–20 nanometers. In addition, intrachain folding produces particles that have great stability compared to other techniques. Control of the precursor chain can also dictate the properties of the nanoparticle, allowing for the production of distinct molecules. There are different varieties of the SCNP method and the type of reaction is dependent on the functional groups involved. However, all the methods benefit more from intramolecular cross-linking rather than intermolecular cross-linking. Homofunctional chain-collapse involves placing a functional group that is likely to bind with itself on the precursor chain and then performing a reaction that couples the functional group. This method often produces particles that are not globular in shape. Instead, heterofunctional coupling is being looked to for improved results. This requires two functional groups which are orthogonally cross-linked. There are many ways to perform the cross-linking in both hetero and homofunctional chain collapses. Data has shown that this method produces nanoparticles with improved spherical shape.

4.8.8. *Protein Particle Composite*

Protein particle composites contain more than one protein or polymer, the addition of which can be used to tune the mechanical and physical properties of a drug delivery vehicle. The method in which they are fabricated depends on the type of particle desired and the differing properties of the additional component. The properties can be a variety of different things such as mechanical properties, electric properties, electromagnetic properties, elasticity, crystallinity, moldability, and many more. When choosing the different type of additional polymer to add, it usually contains an additional characteristic that the main protein does not. Where one protein's structure may be dependent on pH, adding another material to form a stable complex between the two to withstand a lower pH could make more fabrication methods possible. For fabrication of these particles, it depends what end product is desired.

Any previous or following fabrication method that is described can be used to make composite protein-based particle. The only difference between this method and the others is the fact that a protein composite must be made before or after fabrication. For example, if the particles are going to be fabricated using spray drying, a liquid protein mixture can be made before spray drying is done or individual nanoparticles can be formed and then mixed together to create the same product. If a certain percentage of protein is desired in the final product, it is important that an appropriate fabrication method is selected. Fabrication methods can affect the resulting nature of the particle. Particle behavior depends on their surface composition, geometry, and size among other characteristics. There are many more methods that can be utilized for fabricating protein particles apart from what was described in this review.

4.9. Factors to Control Particle Formation

In nanoparticle formation there are many factors that can be controlled to modify drug delivery, such as size, molecular weight, and shape. These factors are mainly determined by the fabrication technique applied but can also be due to the properties of the polymers themselves.

4.9.1. *Size*

Nanoparticle size can vary depending on the molecular weight of the protein polymer used. Typically, nanoparticle size ranges from 1–100 nanometers but they can extend to 1000 nanometers in diameter. One way to control nanoparticle size is to prevent aggregation of the nanoparticles, which can be done by introducing chemicals that help prevent this aggregation by reducing disulfide bonds or by altering the charge state of the polymers. Other factors related to controlling the size of the nanoparticles vary by the technique used to produce them. With the spray drying nanoparticle manufacturing technique, the size of the particles can be altered by changing the size of the nozzle used to spray the polymeric nanoparticle solution into the drying chamber; the size can also be altered by the speed at which the solution is sprayed.

4.9.2. *Shape*

There are many different forms of nanoparticles. The two fundamental types are nanospheres and nanocapsules. The main difference between these types are that nanospheres contain a polymer matrix inside, whereas nanocapsules have a shell that separates the encapsulated polymer from the outside environment. Solid lipid nanospheres are being studied as potential drug carriers due to their matrix morphology. This allows for controlled release and protection of the drug. These particles can be formed by subjecting alkyl cyanoacrylates to polymerization in emulsion. An additional method is precipitating polymers that have already been altered. Solid nanospheres may also be formed using microfluidics method. This method is extremely cost efficient and allows for more control of particle features. The solvent volatility can be altered to shape the surface. Variances in flow rate and the architecture of the devices can create different geometries.

Nanocapsules are somewhat the opposite of solid nanoparticles. This is based on their hydrophobic and hydrophilic interactions. The counter methods can be applied to form nanocapsules. Adding an oil to the emulsion polymerization results in a core-shell formation. Essentially, the presence or absence of oil dictates which type of nanoparticle will form. Another type of nanoparticle is the Janus nanoparticle. These particles consist of two different sides, each with their own properties. These properties can include hydrophobicity and hydrophilicity. The combination of functionalities allows for stimuli response and unique assemblies. Janus particles can be formed by masking. This process involves protecting one region of the particle while the other is functionalized. The mask is then removed, and the final product is a particle with a dual nature particle. Another process is self-assembly. First block copolymers undergo phase separation. Then specific cross-linking must occur, followed by disassembly of larger structures.

4.9.3. *Properties of the Protein*

When designing a protein-based nanoparticles, it is important to consider how the protein will interact with encapsulated drug and physiological environment. Ultimately, a protein with the appropriate molecular weight and IEP must be chosen. The molecular weight of the protein used to create the nanoparticles is important to consider since it can affect how much drug can be effectively stored and the particles targeting mechanism in the body. In some instances a nanoparticles made from a very high or low molecular weight protein can result in lower encapsulation efficiency. A moderate molecular weight protein is often more appropriate and can help to achieve higher encapsulation efficiency. The molecular weight can also contribute to the pathing of the nanoparticle through the body. In addition to molecular weight, the IEP of the protein will affect the stability of the nanoparticle in different environments. At pH near the IEP, nanoparticles may begin to aggregate and decrease in stability. This can inhibit their circulation throughout the body as well as their drug release. Therefore, a protein with the appropriate IEP and molecular weight must be chosen to ensure that nanoparticles withstand certain environments.

CHAPTER V

5. Novel Applications of Protein-Based Nanoparticles

Protein nanoparticles offer a wide range of uses in medicine as both drug delivery vehicles and bioimaging aids.

5.1. Bioimaging

Polymer nanoparticles are gaining traction as contenders to replace typical fluorescent dyes. These are used in non-specific and targeted microscopic imaging. In non-specific imaging, these nanoparticles can be used to dye cells. Data demonstrates that phospholipid encapsulated polymer nanoparticles are successful in providing quality fluorescent imaging of cancer cells. These cells displayed no symptoms of toxicity. In addition, it is possible to tune the wavelength emitted by altering the conjugated protein polymer. These properties, along with an increased circulation period, could lead to applications in vivo.

In addition, protein nanoparticles have a bright future in targeted cellular imaging. These particles have an increased uptake due to the enhanced permeability and retention of advanced tumors. Near IR light can provide excellent imaging quality when paired with a polymer nanoparticle-based probe due to the previously mentioned properties.

Overall, the applications of these particles in the biomedical imaging field are rapidly growing. To enhance the biocompatibility and cellular uptake of nanodiamonds (ND), Khalid et al. encapsulated the material in silk fibroin nanospheres using a co-flow technique. Due to silk fibroin's transparency and low background signal, the photoluminescence of the NDs was not diminished. In fact, NDs encapsulated in silk fibroin spheres fluoresced 2–4 times brighter than NDs alone. The 400 to 600 nm spheres were also found to be highly stabile in an aqueous environment, but began to degrade after one week of incubation at 37 _C. When introduced to fibroblast cells in vitro, the intracellular mobility and diffusion of NDs was improved. Li et al. also used silk fibroin to create nanoparticles for bioimaging, through hydrothermal treatment that simply involved heating the protein at 200∘C for 72 h.

This procedure produced water-soluble, nitrogen-doped, photoluminescent-polymer-like carbonaceous nanospheres (CNSs) that measured approximately 70 nm in diameter and could easily be isolated through filtration. These nanoparticles exhibited low cytotoxicity when incubated with HeLa cells and fluoresced in the perinuclear regions once ingested. CNSs could also be used to image tissue at a depth of 60 to 120 μm with no blinking and low photobleaching.

These studies illustrate the improvements that may result from the incorporation of protein into nanoparticles for bioimaging. In addition to silk fibroin, gelatin nanoparticles have also been utilized as bioimaging platforms. Liu et al. created gelatin nanocapsules containing gold nanoparticles by denaturing gelatin polypeptides that then absorbed onto citrate-stabilized gold nanoparticles.

A thin layer of silica was then used to stabilize these particles that measured approximately 50 nm in diameter and hold promise in Raman-active bioimaging. Gelatin has also been used to coat Cadmium telluride (CdTe) quantum dots (QDs), leading to an improvement in their cytotoxicity and biocompatibility. In this study, Byrne et al. introduced gelatin single- or multi-stranded polypeptides during QD synthesis, to control their growth and nucleation. Functional groups present in the glycine, proline, and 4-hydroxy proline residues of gelatin were then able to interact with the surface of the CdTe QDs, allowing for their coating. When incubated with macrophages, these "jelly dots" were successfully engulfed by the cells, resulting in the illumination of their membranes. When compared to QDs alone, cells exposed to QDs treated with gelatin showed a lower lysosomal pH and cellular permeability, suggesting decreased toxicity of the particles. In both studies, the ability to further functionalize the surface of these particles due to their natural polymer coating,may further enhance their efficacy in bioimaging.

5.2. Drug Delivery Vehicle

Protein-based nanoparticles have also found new use as drug delivery vehicles. In addition to their biocompatibility and biodegradability, the surface of protein nanoparticles can be easily functionalized due to their defined primary structure, while charged proteins can facilitate drug loading through electrostatic interactions. Such particles can also often be fabricated under mild, aqueous conditions, making them easier and safer to process than ones based on synthetic polymers. The use of natural proteins has also been shown to increase cell retention and reduce the effects of toxic byproducts produced during degradation. One such protein used to create nanoparticles for drug delivery is corn zein.

Due to its hydrophobic nature, this protein is especially suited for the prolonged, controlled release of pharmaceuticals. Lai et al. noted this effect when they used the protein to create nanoparticles loaded with the chemotherapeutic agent, 5-Fluorouracil (5-FU). These particles were synthesized using a standard phase separation procedure and measured approximately 115 nm in diameter.

The corn zein particles were able to encapsulate 5-FU at an efficiency of up to 56.7% which was then released after an initial burst of 22.4%. When injected into mice, the nanoparticles remained in circulation for 24 h before localizing to the liver due to their high molecular weight. Corn zein nanoparticles have also been used for the controlled release of vitamin D3, therapeutic proteins such as catalase and superoxide dismutase, and anti-diabetic drugs. In the latter study, Xu et al. developed hollow zein-based nanoparticles through a two-step procedure. This fabrication technique began with the mixing of corn zein and sodium citrate ethanol-based solutions. The zein polymer than aggregated around the sodium citrate crystals, resulting in the formation of particles with a sodium citrate core and zein shell. To create hollow particles, the core-shell particles were added to water, leading to the dissolution of the sodium citrate core.

The resulting hollow nanoparticles measured less than 100 nm in diameter and were able to encapsulate 30% more drug compared to solid zein particles. This drug was then released in a more sustained, prolonged manner over 200 h. When incubated with 3T3 fibroblast cells, the particles were also successfully internalized by cells without effecting their viability.

Other plant-based proteins such as soy protein have also been used to create nanoparticles for the controlled release of nutrients and pharmaceuticals. Due to soy's balanced composition of nonpolar and polar residues, it can act as a versatile carrier by storing drugs with various functional groups. Using a desolvation method and a glutaraldehyde crosslinker, particles measuring between 200 and 300 nm in diameter were fabricated and loaded with curcumin. Curcumin was then released with an initial burst of over 50% within the first 1.5 h, but continued to be released in a more controlled manner over the next 8 h. Although cell studies were not conducted, the established biocompatibility of soy suggests that the particles were act as suitable drug delivery vehicles.

Negatively charged proteins such as keratin have also been used to fabricate nanoparticles that are able to incorporate drugs through electrostatic absorption. When prepared through ionic gelation, keratin particles allowed for the long-term and controlled release of the model drug chlorhexidine (CHX). This fabrication technique involved the dropwise addition of a CHX solution to one containing keratin. Negatively charged carboxylate groups on the outside of keratin aggregates attracted the drug, allowing for its retention. CHX was then gradually released over 140 h in a pH-sensitive manner, with greater release occurring at acidic and neutral pH. Cheng et al. also created keratin-based nanoparticles consisting of varying ratios of the oxidized (keratose or KOS) and reduced (kerateine or KTN) forms of the protein. These particles were created using an ultrasonic dispersion technique and measured between 345 and 400 nm, with decreasing diameter upon the addition of more KOS. The addition of KOS also resulted in an increased release rate of the model drug, Amoxicillin (AMO). The nanoparticles were found to be mucoadhesive due to the electrostatic interaction and disulfide bonding between gastric mucin and KTN, and hydrogen binding with KOS. These results suggest that keratin-based nanoparticles may be an ideal carrier for mucoadhesive drug delivery.

References

[1]   Bowman MC, Ballard TE, Ackerson CJ, Feldheim DL, Margolis D M. J Amer. Chem. Soc. 2008; 130: 6896-7689.

[2]   Mallipeddi R, Rohan LC. Expert opinion on drug delivery 2010; 7: 37-48.

[3]   Taylor U, Klein S, Petersen S, Kues W, Barcikowski S, (2010). Cytomet. Part A. 2010; 77(5): 439-446.

[4]   Lara HH, Ayala-Nuñez NV, Ixtepan-Turrent L, Rodriguez-Padilla C. J nanobiotech. 2010; 8: 1.

[5]   Baram-Pinto D, Shukla S, Perkas N, Gedanken A, Sarid R. Biocon. Chem. 2009; 20: 1497-1502.

[6]   Papp I, Sieben C, Ludwig K, Roskamp M, Böttcher C. (2010) Small 2010; 6(24): 2900-2906.

[7]   Sun L, Singh AK, Vig K, Pillai SR, Singh SR. J Biomed Nanotech. 2008; 4: 149-158.

[8]   Tsai CY, Shiau AL, Chen SY, Chen YH, Cheng PC. Arthritis & Rheumatology 2007; 56(2): 544-554.

[9]   Bhattacharya R, Mukherjee P, Xiong Z, Atala A, Soker S. Nano Letters 4: 2479-2481.

[10]   Arvizo RR, Rana S, Miranda OR, Bhattacharya R, Rotello VM. Nanomed. Nanotech. biol med. 2011; 7(5): 580-587.

[11]   Ramezani N, Ehsanfar Z, Shamsa F, Amin G, Shahverdi HR. Zeitschrift für Natur for schung B 2008; 63: 903-908.

[12]   Chang AL, Khosravi V, Egbert B. J cutaneous pathol. 2006; 33: 809-811.

[13]   Rahman MF, Wang J, Patterson TA, Saini UT, Robinson BL. Toxicol. Let. 2009; 187: 15-21.

[14]   Moss AP, Sugar MD, Hargett MD. Arch Ophthalmol 1979;97: 906- 908.

[15]   Chen W, Zhang Q, Kaplan BL, Baker GL, Kaminski NE. Nanotoxicol. 2014; 8: 11-23.

[16]   Gottlieb NL, Smith PM, Penneys NS, Smith EM. Arthritis & Rheumatol. 1974;17: 56-62.

[17]   Gottlieb NL. Scandin. J Rheumatol. 1983; 12: 10-14.

[18]   Holister P, Weener JW, Vas CR, Harper T, Nanoparticles C. (2003) Technology White Papers nr. 3. Cientifica Nanoporous Materials.

[19]   Feng QL, Wa J, Chen GQ, Cui KZ, Kim TM. J Biomed. Mat. Res B: 2003; 662-668.

[20]   Frens G. Nat phy. sci.1973; 241: 20-22.

[21]   Díaz MR, Vivas Mejia PE. Pharmaceut. 2013; 6(11): 1361-1380.

[22]   Doria G, Conde J, Veigas B, Giestas L, Almeida C. Sensors 2012; 12(2): 1657-1687.

[23]   Qian X, Peng XH, Ansari DO, Yin Goen Q, Chen GZ. Nat. biotech. 2008; 26: 83-90.

[24]   Sperling RA, Parak WJ. Mathematical, Physical and Engineering Sciences 2010; 368: 1333-1383.

[25]   Ghosh P, Han G, De M, Kim CK, Rotello VM (2008) Gold nanoparticles in delivery applications. Advanced drug delivery reviews 60(11): 1307- 1315.

[26]   Nishiyama N (2007) Nanomedicine: nanocarriers shape up for long life. Nature Nanotechnology 2(4): 203-204.

[27]   Chen H, Shao L, Ming T, Sun Z, Zhao C. Small 201; 6(20): 2272-2280.

[28]   Day ES, Morton JG, West JL. J biomech. Engine. 2009; 131(7): 074001.

[29]   Gibson JD, Khanal BP, Zubarev ER. J Ame Chem Soc. 2007; 129(37): 11653-11661.

[30]   Anonymous, 2007a, Agricultural Strategy for Eleventh plan. Some Critical Issues. Govt. of India, pp. 18 (Planning commission).

[31]   Anonymous, 2007b, Eleventh Fire yera plan, 2007-2012, Agriculture, Rural Development, Industry, Services and physical Infrastructure, pp, 1-40.

[32]   Kalpana Sastry, R., Rashmi, H. B., Rao N. H., Ilyas, S. M., 2009, Nanotechnology and agriculture in India: the second green revolution? Invited paper presented in: Session IV. OECD Conference on Potential Environmental Beneûts of Nanotechnology: Fostering Safe Innovation-Led Growth, 15-17 July, OECD Conference Centre, Paris, France.

[33]   Kalpana Sastry, R., Rashmi, H. B., Rao, N. H. J. Intell. Prop. Rights, 2010; 1(5): 197-205.

[34]   Li, Z. Z., Chen, J. F., Liu, F., Liu, A. Q., Wang, Q. and Sun, H. Y., 2007, Study of UV-shielding properties of novel porous hollow silica nanoparticle carriers for avermectin. Pest Management Sci., 63:241-246.

[35]   Li, Z. Z., Chen, J. F., Liu, F., Liu, A. Q., Wang, Q. and Sun, H. Y. Pest Management Sci.,2007; 63:241-246.

[36]   Frederiksen, H.K., Kristensen, H. G. and Pedersen, M. J. Control Release, 2003; 86: 243-252.

[37]   Liu, Y., Tong, Z. and Prudhomme, R. K. Pest Management Sci., 64: 808-812.

[38]   Barik, T. K., Sahu, B. and Swain, V., 2008, Nanosilica-from medicine to pest control. Parasitol Res.,103: 253-258.

[39]   Lai, F., Wissing, S. A., Müller, R. H. and Fadda, A. M. AAPS Pharm. Sci. Tech., 2006; 7: 1-9.

[40]   Vandergheynst, J., Scher, H., Guo Hy, Schultz, D. Bio Control, 2007; 52: 207-29.

[41]   Ghormade, V., Deshpande, M. V. and Panikar, K. M. Biotechnol. Adv., 2011; 29: 792-803.

[42]   Wilson, M. A., Tran, N. H., Milev, A. S., Kannangara, G. S. K. Geoderma; 2008; 146: 291-302.

[43]   Torney F, Trewyn BG, Lin VS, Wang K. Nature Nanotechnol. 2007; 2: 295-300.

[44]   Vijayakumar, P. S., Abhilash, O. U., Khan, B.M. and Prasad, B. L. V. Adv. Funct. Mater. 2010; 20: 2416-2423.

[45]   Zhang, X., Zhang, J. and Zhu, K. Y. Insect Mol. Biol., 2010; 19: 683-693.

[46]   Guo, Y. R., Liu, S. H., Gui, W. J. and Zhu, G. N. Anal. Biochem., 2009; 389: 32-39

[47]   Lisa, M., Chouhan, R. S., Vinayaka, A. C., Manonmani, H. K. and Thakur, M. S. Biosens. Bioelectron., 2009; 25: 224-227.

[48]   Guan, H., Chi, D., Yu, J. and Li, X. Biochem Physiol., 2008; 92: 83-91.

[49]   Wang, H., Wang, J., Choi, D., Tang, Z., Wu, H. and Lin, Y. Biosens. Bioelectron. 2009; 24: 2377-83.

[50]   Ramanathan, M., Luckarift, H. R., Sarsenova, A., Wild, J. R., Ramanculov, E. R. and Olse, E. V. Biointer. 2009; 73: 58-64.

[51]   Kaushik, A., Solanki, P. R., Ansarib, A. A., Malhotra, B. D. and Ahmad, S. Biochem. Engineer J., 2009; 46: 132-40.

[52]   Paknikar, K. M., Nagpal, V., Pethkar, A. V. and Rajwade, J. M. Sci. Tech. Adv. Mat., 2005; 6:370–374.

[53]   Guan, H., Chi, D., Yu, J. and Li, X. Biochem Physiol., 2008; 92: 83-91.

[54]   Hef fer, P. and Prud'homme, M., 2012, Fertilizer outlook 2012-2016. Paper presented at the 80th IFA annual conference, 21-23 May, Doha (Qatar).

[55]   Tilman, D., Knops, J., Wedin, D. and Reich, P. Oxford University Press, Oxford, pp, 2002; 21-35.

[56]   Mahmoudi, M. Lynch, I. Ejtehadi, M.R. Monopoli, M.P. Bombelli, F.B. Laurent, S. Protein-nanoparticle interactions: Chem. Rev. 2011; 111, 5610–5637.

[57]   Weber, C., Coester, C., Kreuter, J., Langer, K. Int. J. Pharm. 2000; 194, 91–102.

[58]   Wongpinyochit, T. Uhlmann, P. Urquhart, A.J. Seib, F.P. Biomacromol. 2015; 16, 3712–3722.

[59]   Zhi, X. Wang, Y. Li, P. Yuan, J. Shen, J. RSC Adv. 2015; 5, 82334–82341.

[60]   Posadas, I. Monteagudo, S. Ceña, V. Nanomed. 2016; 11, 833–849.

[61]   Bajpai, A. Choubey, J. J. Mater. Sci. Mater. Med. 2006; 17, 345–358.

[62]   Herrero-Vanrell, R. Rincon, A. Alonso, M. Reboto, V. Molina-Martinez, I. Rodriguez-Cabello, J. J. Control. Release 2005; 102, 113–122.

[63]   Hurtado-López, P. Murdan, S. J. Microencapsul. 2006; 23, 303–314.

[64]   Teng, Z. Luo, Y. Wang, Q. J. Agric. Food Chem. 2012; 60, 2712–2720.

[65]   Saralidze, K. Koole, L.H. Knetsch, M.L. Mat. 2010; 3, 3537–3564.

[66]   Paciello, A. Amalfitano, G. Garziano, A. Urciuolo, F. Netti, PA. Adv. Healthc. Mater. 2016; 5, 2655–2666.

[67]   Chen, X.; Lv, G.; Zhang, J.; Tang, S.; Yan, Y.; Wu, Z.; Su, J.; Wei, J. Int. J. Nanomed. 2014.